The Riddled Chain

The
Riddled Chain

Chance, Coincidence, and Chaos
in Human Evolution

JEFFREY K. McKEE

RUTGERS UNIVERSITY PRESS
New Brunswick, New Jersey, and London

Library of Congress Cataloging-in-Publication Data
McKee, Jeffrey Kevin.
    The riddled chain : chance, coincidence, and chaos in human evolution /
    Jeffrey K. McKee.
        p.   cm.
    Includes bibliographical references and index.
    ISBN 0-8135-2783-X (alk. paper)
    1. Human evolution.   2. Chance.
GN281.M396   2000
599.93'8—dc21   99-045720

British Cataloging-in-Publication data for this book is available
from the British Library

Manufactured in the United States of America

For Jean, Nathan, and Aaron,
each of whom inspires me every day

Contents

Preface

THROUGHOUT HIS BRILLIANT academic career, Thomas Henry Huxley not only pursued rigorous science but also made science accessible to the layperson through his lectures and essays. His work remains to this day one of the best introductions to the evolutionary theory of Charles Darwin. Nearly a century after Huxley's death, I began writing this book to honor his scientific insight as well as to try and capture his spirit in educating the public about the nature of our evolution.

The Riddled Chain is as much about the scientific process as it is about the mechanisms of human evolution. Its raw materials come from fossil excavations in the field, observations in labs, calculations on computers, and the ruminations of many clever minds. I weave these elements into what can only be a personal view. In so doing I liberally use my own work where it can be instrumental in making a point, partly due to my comfort with the subject matter and partly in an attempt to give the reader a feel for the scientific process. Nevertheless this book is about general principles, not my own research, and I aim to tweak the reader's imagination toward new ideas.

I believe that science is fun, and that the joys (and frustrations) of research should be reflected in the way we scientists portray our work. Thus this book is written in a spirited and nontechnical manner, despite its serious content. I hope that my academic colleagues will enjoy a wry grin while thinking seriously about the issues discussed.

Beyond Thomas Huxley and Charles Darwin, I have been inspired by many scholars. Only some of these are cited, and even some of the greats such as Ernst Mayr or Sewall Wright are mentioned only in passing, despite their strong influences on me. I owe a huge debt of gratitude to all of them.

I am indebted to many for my field experiences. Phillip Tobias gave me the opportunity to work at Taung and graciously shared his keen insight over a number of years. Alun Hughes taught me how to dig through rock and be a "fossil farmer." The people of Buxton allowed me to make a home in their village, always making me feel welcome. I am particularly grateful to all those who worked for me at the Taung and Makapansgat fossil sites and who demonstrated the greatest spirit of humanity. My colleagues in the field were always of help, namely Michel Toussaint and Kevin Kuykendall. Likewise the students who traveled with me to Taung and Makapansgat gave me their labor, attention, and friendship. Graduate students who helped to shape the excavations as assistant field directors were very important; thanks to Wallace Scott, Lee Berger, Patrick Quinney, and Colin Menter.

I should also like to make special mention of Bob Brain, a rare breed of scientist who calmly blends meticulous work with precious insight, all the while enjoying and involving his family. Thanks, Bob, for the inspiration on how to live a scientific life.

Although the title page of this book has me as the sole author, many people have taken part in the writing. Comments on early drafts helped reform the book from its scratchy beginnings, and careful editing of later drafts led to the product in your hands. Thus I gratefully acknowledge, in alphabetical order, Sue Armstrong, Ellen Bartlett, Jarrod Burks, Jack Harris, Will Hively, Helen Hsu, Kirk Jensen, Alan Mann, Bill McKee, Jean McKee, Patty Parker, Gene Poirier, Alan Walker, and a number of anonymous referees.

Many thanks, again, to Helen Hsu, acquisitions editor at Rutgers University Press, who skillfully and kindly ushered this book to completion.

And, finally, many thanks to my family for their love and support throughout this project.

The Riddled Chain

1

Chance, Coincidence, and Chaos

THE EVOLUTION of life on earth has produced some exquisite and peculiar beasts, among which humans are a distinct curiosity. We tend to set ourselves apart from other animals, for sentient beings hardly seem to fit into nature's intricate order of life. But humans are really no less marvelous than other living species with whom we share a long and complex evolutionary history. All existing life forms endured a struggle for existence and evolved seemingly inspired adaptations. We are just another product of nature, yet we humans uniquely stride on two legs, carrying aloft a curious brain with which we can probe our origins and dare to foresee the future.

Our very nature makes us curious about how and why we evolved, but the particulars of our evolutionary creation have been somewhat elusive to traditional science. We understand some key principles, rooted in Darwinian biology, but sometimes the simple means of evolutionary progress do not seem to offer a full explanation of our biological gifts. How could an aimless evolutionary process, patching together random biological novelties and oddities through trial and error, lead to *Homo sapiens*?

Much of our difficulty in probing our origins is due to the tremendous time scale required for the process of evolution to unfold, something that is not easy to comprehend. Only a few sparse fossils from the distant past, and keen observations of life on earth today, can spur the scientific imagination toward testable hypotheses of how evolution led to the menagerie populating the planet today—and how we were afforded the luxury of questioning our place in nature. Yet even with the fragmentary fossil evidence we have, our understanding is limited solely by our imaginations.

One can only dream of the grandest biological experiment: restarting the earth at the inception of life and watching the evolutionary story play itself out. One wonders how our history and prehistory might differ if we were to start again.[1] Would birds and baboons alike ultimately find a place on the planet after billions of years of evolution from the simplest of living organisms? Would our place in nature be the same, or would humans even have any place in nature at all? Most likely not. Initial life forms would have evolved into an equally astonishing array of creatures, to be sure, being aimlessly self-propelled through a tortuous evolutionary history with no particular destiny. It is unlikely that life was destined to have included us. This is because the evolution of life is subject to fates wantonly dictated by three ubiquitous and mischievous forces: chance, coincidence, and chaos.

Chance

I really should not be here. It is not that I have no right to be here in Ohio or anywhere else, but that I *probably* should not be anywhere at all. Biological principles made me what I am, but my good fortune in being present on earth is the highly unlikely consequence of a long series of chance events. Indeed, chance itself may be considered as a key biological principle.

Chance played more of a role in my immediate origin than I should care to admit. When I was born, the youngest of four boys, it was largely because my mother wanted a daughter. By having four children the odds for at least one daughter were on my mother's side, but the less likely outcome of the birth stakes played a little trick on her. Lucky me, for otherwise I would not have been conceived.

So how did all this happen? Let's assume, for convenience' sake, the chances of having a boy or a girl at each birth are exactly even—a 50 percent chance for the birth of Susan Gayle McKee. (Yes, her name had been chosen.) Had the firstborn of my family chanced upon an X chromosome from my father, and thus become a girl, I never would have been conceived years later. But my mother's peculiar saga had begun, much to my ultimate benefit. It was a Y chromosome that started a chain of events by producing a boy.

The second child also could have been a girl. The laws of probability tell us that chance is multiplicative, so the chance of my poor mother having two boys in a row was only 25 percent. We understand in retrospect that the third son was not exactly planned, another chance event, but to keep things simple, he too had a 50 percent chance of being the daughter my mother so desired. Multiply him in, and there was a 12.5 percent chance, or just one chance in eight, that there should have been any need for my conception.

Just a different gleam from my father's eye, and I too could have been Susan Gayle McKee, perhaps a distinguished suburban lawyer rather than an itinerant anthropologist.[2] My mother would have been pleased, but it was not to be. Taking matters back further, now that we are thinking along these lines, none of the McKee boys would be here at all if my father had not been born a boy, and my mother born a girl. Put all these chance events of two generations together and there is less than a 1.6 percent probability, merely one chance out of sixty-four, that I should be here on earth. Keep in mind that I am considering only gender selection here. The fragility of the human developmental process, along with a myriad of biological and social factors, makes me even more unlikely.

Yet here I am, digging up ancient fossils and studying the remote human past, despite the improbability of *my* recent past. To what can we attribute this sequence of events? Some would call it fate; others may invoke manifest destiny. I prefer to look at it more scientifically, in which case, at least statistically speaking, the events leading to my birth were a culmination of nothing more than dumb luck—a product of chance. These things happen.

Some may look at my analysis of my own improbability and quibble with my mathematics. Fair enough, for they may look at things from a population perspective: it was *likely* that a boy such as myself would have been born, and *probable* that some families would consist of four unruly boys. True, but I cannot help thinking that if time were rolled back and the sequence started over, with just one tiny difference in the fertilization of any egg that became a member of the McKee family, I would not be here and you would not be reading this book.

As we go back generation after generation, the probabilities of coming up with exactly me become diminishingly small. Certainly my ancestors hoped that their lineage would carry on, but the precise way in which things happened would have been well beyond their prognostication. One chance event led to another and, for better or worse, here I am.

Certainly you must realize that you are equally unlikely. The same laws of probability apply to your family as well as mine. The same genetic principles determining gender and a host of other characteristics have been filtered through the generations to come up with something that is uniquely you, unless of course you have an identical twin. You are lucky to be here, for a plethora of scenarios could have played out over the years. Every human being, as well as every monkey, bird, and insect, is the uniquely improbable result of past combinations of genetic material. It is that specific improbability of every individual that makes each of us unique, and gives evolution something to work with.

Not only is the probability of your own individual existence very tenuous, but your species—our species, *Homo sapiens*—is also a fluke of familial descent. We are lucky to be here, and we are exceptionally lucky to be able to think and read and write about our good fortune. Baboons, pigeons, and cave crickets were not afforded the same modicum of cerebral capabilities, no matter how unique and well-adapted they may be. And they were bred by the same basic principles that led to us. We are just, as Thomas Henry Huxley described us in 1854, an "aberrant modification" of an evolutionary theme.[3]

Now this may be a bit disconcerting to some people. Sentient beings, sapient *human* beings, have always thought that there was something inevitable about them. Even devout Darwinian evolutionists tend to put our own immodest species at the top branch of the evolutionary tree, as if we were somewhat better and more evolved than other living species. But, as we shall see throughout this book, chance has played a role in putting every living thing at the top of the present evolutionary tree.

Some have looked with discomfort at the unlikelihood of the evolution of a large-brained, intelligent species and proclaimed evolutionary theory a farce. Statistically speaking, our evolution seems to be too improbable even to consider. Our evolutionary probability is indeed minuscule. But such arguments invoke the same logical twist that I used above to claim my own improbability. Any specific product of biological evolution, be it human or giraffe, is unlikely. What is likely is that *some* sort of species would have descended from our own remote ancestors. Just as contemporary statisticians could have predicted that a boy such as myself would be born in Ohio in 1958, a prehistoric prognosticator could have foreseen some sort of peculiar species emerging from the animals then on earth. The boy did not have to be me, and the species did not have to be us. It was a matter of chance, intermingled with other principles of evolutionary biology.

Chance is one of the most horrifying words in the lexicon of a scientist. It strikes terror into our hearts, for we want rational, principled explanations of the world as we know it. Scientists try to package the phenomena of physics and biology alike into neat little explanatory boxes driven by deterministic processes in which chance plays no role. This is a pity, because by embracing the powers of chance, rather than balking at the use of the word, scientists may find that some very interesting things can happen by chance, including the evolution of the human species. We may even discover the keys to what *increased* our chances of surviving and evolving.

So how did all this happen? How did a series of chance events lead to complex sentient beings? Charles Darwin gave us the answer, or at least a large part of the answer, about a hundred years before one of his adamant supporters chanced to be born in Ohio. He called the directional force of evolution "natural selection"—the way nature "selects" among the variants born of chance. In the words of Sir Ronald Fisher, "natural selection is a mechanism for generating an exceedingly high degree of improbability."[4]

Natural selection can shape and mold populations by allowing the fittest individuals to survive and reproduce. Among those individuals, in the last 3 or 4 million years, were our own ancestors as well as the ancestors of every creature on earth today. Our ancestors made it through natural selection. What luck.

But that luck may have been shaped by a series of timely coincidences.

Coincidence

The events of one's life, or of many lives, often coincide in ways that invite the human imagination to conjure up intriguing but tenuous meanings to simultaneous circumstances. *Coincidence* plays a tremendous role in the way we perceive and interpret our universe. But the meaning of coincidence can be viewed in many different lights.

Many people firmly believe that tragic events, particularly deaths, always occur in threes. Thus, once two friends or relatives have passed away, such believers in this rule anxiously await the third demise. It is truly amazing how often we can observe such packages of three coincidental misfortunes, but more often than not the package is an artificial product of our minds' attempts to rationalize what we see. People die every day, and other things are always going wrong, so it is easy to choose three people we know or three personally significant circumstances to fulfill our own prophecies. Such sequences of events are thus *mere* coincidence, with no greater meaning.

Sometimes, however, events coincide for distinct reasons. One circumstance may cause another, either directly or indirectly. For example, your purchase of this book may cause my income to rise, and I am truly grateful for that direct effect, as is my growing family. More indirectly, what I am writing today may cause you to view life differently, a coincidence that is my primary purpose for writing this book. But up to this point, our lives have been largely independent.

Likewise, distinct events may share the same cause. The end of a war may cause an increase in the birthrate in many families; for this coincidence I hold a measured amount of gratitude, as my birth was at the tail end of the baby boom that followed World War II. But it is not always easy to differentiate between mere coincidences, caused by chance or perceived artificially, and significant coincidences that stem from recognizable causes.

It is no mere coincidence that I began writing this particular section of my book on the twelfth day of February. This day coincides with a particular anniversary, in fact two anniversaries, and my choice to write was made with forethought and eager anticipation. On February 12, 1809, Charles Darwin, the great naturalist and founder of modern biological thought, was born in Shrewsbury, England. On this day, also in 1809, Abraham Lincoln, the great American statesman and sixteenth president of the United States, was born in Hardin County, Kentucky. Both Darwin and Lincoln wrote words that have found their way into my personal library, and indeed my library contains biographies of both. The words and deeds of Darwin and Lincoln, two unrelated people born an ocean apart on the *very same day*, shaped human life and thought on earth for well over a century, and will continue to do so. What a coincidence!

Charles Darwin and Abraham Lincoln shared more than just birthdays, greatness, and long-term impact on human thought. Both lost their mothers at an early age, Darwin in 1817 and Lincoln in 1818. In 1831 Lincoln began life on his own as Darwin set sail on the *Beagle*, also leaving his former family life behind. Darwin published his first paper in 1836, a rite of passage for any academic, as Lincoln paralleled such an achievement by attaining his license to practice law. The following year, Lincoln moved to Springfield, Illinois, and presaged his future career with his first public statement on slavery; Darwin moved to London and began his all-important notebook on the transmutation of species, inscribing thoughts that soon led to the concept of evolution through natural selection.

The year 1858 was particularly significant for both Darwin and Lincoln. Darwin announced the theory of evolution through natural selection, beginning a debate on the nature of our biological origins that has lasted to this day. The following month, Abraham Lincoln held the first of his famous debates with Stephen Douglas on the issue of human slavery, changing forever the course of political events and human race relations.

From 1858, the lives of Abraham Lincoln and Charles Darwin diverged. Lincoln became the president of the United States in 1861, fought for the preservation of the Union and the emancipation of slaves, and briefly knew victory before his life was cut tragically short by an assassin in 1865. Darwin, on the other hand, lived a fruitful and longer but not healthful life, publishing significant works until his death in 1882. Both did die for their convictions: Lincoln for being perceived as a tyrant by John Wilkes Booth, and Darwin, so it is said, from a slow nervous illness brought on by his dedication to an unpopular view of life. However one looks at it, none can deny that besides the coincidence of their birth, the lives of Darwin and Lincoln contained many significant parallels. Or can it be denied?

One way to test for the possible significance of coincidence, such as that of Darwin and Lincoln, is to look for a common cause. Certainly the social influences of the nineteenth century shaped the courses of their lives. The circumstance of one human enslaving another went across the grain of Lincoln's thought, as it did Darwin's, and catalyzed a heartfelt and eventually effective reaction. But Lincoln had to work against established opinion; so the immediate cause of Lincoln's approach to life was not clear. Likewise, in the England Darwin knew, the works of geologists and economists influenced Darwin's interpretations of the living and fossilized life forms he so keenly observed. But his deductions of evolution through natural selection flew in the face of ensconced Victorian thought. So whereas they were influenced strongly by their times, both Lincoln and Darwin saw through the indoctrinated views of life and strove toward new light. Perhaps their strong-willed independence was more than mere coincidence, with a very specific and principled cause.

For a possible explanation of coincidence, one can turn, as people often do, to the influence of the stars and planets. As an amateur but enchanted stargazer, I marvel that my computer can take me to any place and any time and exhibit the sky in its full glory on a small screen above my keyboard. According to the computer program, the sky on February 12 was a sight to behold in 1809, from England or Kentucky. On the day on which Lincoln and Darwin joined the ranks of the living, Mercury, the messenger of the gods, was in their zodiac birth sign of Aquarius. Maybe this foretold the impact their messages would have. Indeed, Mars, the god of war, was in the constellation Virgo, the virgin; astrology is neither my forte nor interest, but one could interpret the sign in terms of the battles both men were to fight upon breaking fresh new ground. I have no idea of the possible significance of Uranus abutting Libra on that particular night, but certainly some astrologer could enlighten us as to the mystic significance of the event.

Such astronomical observances, or astrological fantasies I suspect, reach much too far for an explanation. The point is that one finds coincidence when one looks for it, and the human mind constantly searches for just such correlations. This habit of searching for patterns is ingrained in our nature for good reason—our curiosity carved the path of our evolution. But our ruminations may not be of any significance; the coincidences may be real, but they are not necessarily important. One must neither abandon principled rules of causation nor ignore the viewpoint that comes from greater circumspection. To put it into perspective, by some very rough calculations I estimate that around ninety thousand people were born on February 12, 1809. Despite the same planetary influences across the globe, only two, Charles Darwin and Abraham Lincoln, led lives that still resonate consciously in our thoughts today.

The critique of coincidence can be taken further. I chose the parallel dates of Darwin's and Lincoln's significant life events at the expense of other, possibly less interesting, noncoincidental occurrences. Poor Lincoln was rejected in his 1836 proposal of marriage to Mary Owen; Darwin married *and* had a first child in 1840, beating Lincoln to betrothal by well over two years and to fatherhood by three. Such discrepancies are easily overlooked in a well-played game of coincidence-seeking.

But a game it is. My entirely subjective analysis of Darwin and Lincoln is reminiscent of an intellectual game imagined by Hermann Hesse in his classic 1943 novel, *The Glass Bead Game.* As Hesse wrote: "Then there were the tempestuous letters from abstruse experimenters who could arrive at the most astounding conclusions from, say, a comparison of the horoscope of Goethe and Spinoza; such letters often included pretty and seemingly enlightening geometric drawings in several colors."[5] Both scientists and pseudoscientists alike fall into the same playful habits of searching for and finding coincidence. It is, after all, good intellectual fun.

Coincidence, be it by cause or caprice, is part of all lives. Interpretations of coincidence can be fraught with error, but they are worth finding when a true, verifiable causal agent can be established. The distant past, heralded by the fossil record, is as subject to the dangers inherent in the overinterpretation of coincidence as it is to the rewards of discovering true causal agents of coincidental change. We may find from the sparse fossil evidence that one species arises as another declines, or even that evolutionary changes among many species appear to have coincided. A good scientific mind is then inclined to propose that such apparently simultaneous events had a common cause. After all, cause and effect in evolution is one of the processes we want to grasp. The job of a scientist is then to test the hypothesis of cause and effect by looking at the fossil record more closely and circumspectly. As it turns out, discerning the difference between mere coincidence and important causation is not only difficult but is severely hampered by the fragmentary nature of the fossil record.

The fossil record we extract from the ground is a notoriously incomplete sampling of past life. Fossilization and preservation of any individual animal is largely a matter of chance, using *chance* in the sense already explained. If the skeletons and imprints of every living being from the past had been fossilized and preserved intact, we would be walking upon a ground so full of incredible fossils that it would be an embarrassment of riches for even the most ambitious and assiduous paleontologists. But most plants and animals die and disintegrate without leaving a recognizable trace for posterity. Once the remains of a life are preserved, subsequent discovery of the resultant fossils is also largely a product of chance, as we shall see.

Our biographical information on the lives and deaths of ancestral forms is thus a mere sketch, and sometimes as lacking as for the ninety thousand or so humans born on February 12, 1809.[6] Had the life histories of Darwin and Lincoln been subjected to the vagaries that characterize fossilization, we might not have been able to note the incredible coincidence of their simultaneous births. They would have appeared in the historical record at roughly the same time, but there would have been a lack of precision with which we could have identified their births, and the potential significance of February 12 would have been glossed over.

On the other hand, if historical records were like fossil records, births that coincided somewhat less precisely may have produced the illusion of simultaneity. Thus, Carolus Linnaeus, the Swedish botanist who founded our taxonomic system of naming and categorizing life on earth, born in 1707, would appear to have joined the human family at roughly the same time as Benjamin Franklin, the famous American statesman and philosopher who was born in 1706. Likewise, the birth in 1744 of Jean-Baptiste de Lamarck, an important French naturalist and early evolutionist, would seem to coincide with that of Thomas Jefferson, yet another American statesman of great historical significance, born but a year earlier in 1743.

One can be remarkably successful in roughly correlating the births of significant players in the history of European biological thought with those of great American statesmen. I tired of this little game after getting through only half of the alphabet, but I did happen to note that Erasmus Darwin, evolutionist and (biological) grandfather of Charles Darwin, and George Washington, statesman and (nonbiological) "father" of the United States of America, were born in 1731 and 1732 respectively.

My impatience with pursuing such historical coincidences and near coincidences stemmed from my firm belief that they hold no significance beyond the pleasures of gaming. I find no useful notions stemming from roughly simultaneous birthdays of biologists on one side of the Atlantic and statesmen on the other. Yet there may have been, and probably were, significant influences guiding each notable individual toward great thoughts and deeds. Such influences and perhaps general principles should not be lost to academic inquiry. Indeed, it is our scientific duty to test coincidences for possible meaning.

Just as it is easy to go astray with an imprecise or forced history, so it is with the fossil record. Paleontologists and evolutionary biologists must be wary of coincidences that are not real, such as close but not quite simultaneous origins or extinctions of different species. The simultaneous origin of a swifter impala with that of a more nimble cat to pursue it may be a clue of great evolutionary significance. Or it may be nothing but *mere* coincidence. Discerning the difference between mere coincidence and those events which reflect true cause and effect is one of the greatest challenges facing evolutionary biologists, and particularly paleontologists. Where true coincidences occur they must not be ignored; correlations between the expansion of the ancestral human brain and the appearance of stone tools, or even between changes in the brain and other parts of the body, *could* be important. The very essence of human success may have culminated from a fortunate set of significant coincidences. Scientists are constantly searching for clues to such coincidental events of true evolutionary significance.

Chaos

In an incomplete fossil record, a cause of coincidental effects (or of any effect at all) may or may not be perceived. This can get frustrating. What if the cause, or at least a partial cause, of a significant evolutionary event was something as seemingly insignificant as a butterfly's passing by? A large carnivorous cat, destined for a successful kill of a complacent impala grazing upwind, may snap fruitlessly at a passing butterfly (as our pet cats and dogs often do). The action alarms the nearby animals, and the cat thus loses the chance to find substantive nutrition in the now escaped herd of impalas. For the lack of nutrition, the carnivore dies. Its potential offspring are never born, and grandcubs become an impossibility. Eventually another species is lost, not solely to caprice but specifically because of a passing butterfly. To an evolutionary biologist, as to a mathematician, such a sequence of events represents *chaos*. And it happens time and time again.

To understand chaos in the context of evolution, we must first define what chaos is, and fit the concept in with chance and coincidence. Chaos has many implications. To most people it is the opposite of order. Chaos may mean anarchy to a politician or unpredictability to a mathematician. It is often a cause for despair. But do not despair, for chaos is one of the reasons you are here. Indeed, in Greek mythology, Chaos is the yawning void from which all things come. Chaos can lead to order, if the chaotic system chances upon a lucky coincidence. But we must back up a bit to understand such nuances.

Unpredictability is at the heart of chaos theory, a branch of mathematics. Chaotic systems can be represented by the weather, which is notoriously unpredictable, or evolution through natural selection. Both systems rely upon very specific principles, such as the exact temperature and relative humidity that produce a rain cloud or the precise genes and environmental interactions that make up the lives of evolving animals subjected to natural selection. But there are many of these principles acting at once.

Chaos is derived from precise principles, exacting causes and effects. If precisely the same conditions start a chaotic system in motion, the result will be the same every time. That is the point: one must start with *precisely* the same conditions down to the most minute details. For the same storm to ensue more than once, all the weather conditions to the finest degree of temperature and the right accumulation of moisture in the air must exist. For the same species to evolve, every link in the chain must be precisely the same, down to the level of the genes of individuals in an initial population.

Chaotic systems are extremely sensitive to initial conditions. At a very fine level of analysis, chaos explains chance. If you were to go to a casino and roll a pair of dice from your hand, the result would appear to be the result of chance. But just imagine if you could control the roll down to the finest detail. With the dice accurately positioned in your hands, an exacting back swing of the forearm, a perfectly measured thrust forward, and precise timing of release at an optimally controlled angle, you could roll the same every time. You would be rich, and the casino would go broke. But you do not have that level of control, no matter how hard you try. On the level of control at which our nerves, muscles, and senses operate, the far more precise rules governing the physics of rolling dice come together in a chaotic fashion, and unpredictable manner, to result in the appearance of chance. Chance is thus the result of chaos, not vice versa.

Let's go back to the improbability of my birth. I attributed my familial sequence of male births to chance, but at a finer level of analysis the appearance of chance was the result of chaos. A very specific sperm had to follow a particular course toward an individual egg at precisely the right time. One unmentionable change of even the slightest proportion and you would not be reading this book today. Like rolling the dice, it was a matter of physical precision that resulted in me rather than anybody else. It was chaos, governed by laws of physics and biology but resulting in something unique every time. Perhaps these laws can provide some solace to those who fear the uncontrollable lawlessness implied by chance. As the great paleontologist William K. Gregory stated in 1949, "Chance should not be contrasted or logically opposed to Law, but both are merely different aspects of one continuous reality."[7]

Many scientists have taken up the concept or concepts of chaos and shaped a definition to suit themselves. Here the term *chaos* will be restricted to its currently applied meaning: chaos represents unpredictability based on sensitivity to initial conditions. Historical contingencies of even the smallest proportion may eventually have a profound effect, even if every event is controlled by precise rules and principles.

Unpredictability is an easy concept. We do not know what will happen next, even with simple principles at work. For example, my grandfather, who presumably understood the basic mechanisms of human reproduction, could not have predicted that his son would have four boys. Likewise, as we shall see, the exact, long-term course of evolution is also unpredictable. One need not impose chance events on evolution to result in unpredictability. The impact of asteroids, for instance, may have skewed the course of evolution in unpredictable directions, but its course would have been unpredictable even if no asteroids ever disrupted an evolutionary trajectory. Evolving systems undergoing natural selection are so strongly affected by the nature of the plants and animals present at the beginning of any time we consider that they are intrinsically chaotic. Evolution depends on initial conditions.

Such extreme sensitivity to initial conditions is best illustrated by a hypothetical event called the "butterfly effect." Some people collect butterflies as a hobby; I collect butterfly effects, and you can view them at your leisure in the box below.[8] The butterfly effect was originally conceived by Edward Lorenz, a researcher of meteorology, who saw the importance of initial conditions in trying to predict the weather. Thus, as you can see from the collection, most butterfly effects involve the weather in North America being changed by one or more butterflies in either South America or Asia. But one can have much more fun with the butterfly effect.

• A Butterfly Collection •

"Does the flap of a butterfly's wings in Brazil set off a tornado in Texas?"

—*Edward Lorenz*

"The flapping of a single butterfly's wing today produces a tiny change in the state of the atmosphere. Over a period of time, what the atmosphere actually does diverges from what it would have done. So, in a month's time, a tornado that would have devastated the Indonesian coast doesn't happen. Or maybe one that wasn't going to happen, does."

—*Ian Stewart*

". . . the notion that a butterfly stirring the air today in Peking can transform storm systems next month in New York."

—*James Gleick*

". . . a butterfly flaps its wings over the Amazon rain forest, and sets in motion events that lead to a storm in Chicago."

—*Roger Lewin*

". . . a butterfly in the Amazon might, in principle, ultimately alter the weather in Kansas. Dorothy's trip to Oz rode on tiny wings."

—*Stuart Kauffman*

"The beating of a butterfly's wings in China can affect the course of an Atlantic hurricane."

—*Douglas Adams*

As an evolutionist, my idea of the butterfly effect is that described earlier, where the butterfly distracts a cat from its prey, condemning the carnivorous species to extinction. One can imagine a wealth of such chaotic foibles in the system. Even with the butterfly distraction, a more tenacious cat or dog in the wild may successfully catch its prey, live to reproduce, and pass on its genes of successful gazelle-hunting and butterfly-ignoring; in turn, this may lead to the demise of its prey, its food source, and ultimately but belatedly to the extinction of the carnivorous beast itself. That damned butterfly can have just about any effect, or lack thereof, and thus makes long-term prediction impossible.

More relevant to our subject of study, a single lost child who stumbled into a dark watery grave changed the course of science forever, nearly 3 million years after his premature death. The chance fossilization of the child's skull set the initial conditions for many of our concepts of human evolution. Indeed, butterfly effects are relevant not only to the evolution of life but also to the development of our interpretations of how life evolved.

Like coincidences—and the fossilization of a child's skull is nothing more than a chance coincidence—initial conditions and their unpredictably chaotic effects pose a problem to scientists who deal with the past. Sometimes when I talk about chaos and explain how it could have shaped human evolution, my colleagues toss their arms up in despair and ask why it is worth studying anything in evolution if some stupid butterfly could be the key to understanding everything. It gets worse when I include the role of chance and add insult to injury by pointing out that much of what we see in the fossil record is mere coincidence. But there is more to understanding chaos. And there is more to understanding evolution.

One need not despair at all. It is simply a matter of asking the right scientific questions. Although the initial conditions may help lead to the evolution or extinction of an animal, there still must be basic principles at work, and such principles are what scientists are after. Not everything can be attributed to the whims of a butterfly. The butterfly effect is simply like that of planting a seed. Sometimes the seed will grow into a tree, later to be used to build a house or to fuel a fire, whereas other times the seed will not germinate at all. At work are basic biological principles of plant reproduction and growth, and if a tree grows to be utilized, basic principles of architecture or combustion are still key considerations.

Trends in the evolution of animals also emerge from simple principles. Species origins and extinctions follow intriguing patterns, despite their chaotic unpredictability. Basic Darwinian principles such as natural selection, as well as ecological norms involving food supply and demand, shape and drive evolution. It is worth understanding those principles and how they apply to the evolution of a species such as our own. Yet there is also value in understanding evolution in the context of chaos theory, for as we shall see, chaos develops when the various forces of evolution combine. Furthermore, the initial conditions of the human form played a large role in determining the eventual outcome we know as *Homo sapiens.*

But if the evolution of life, like the course of an Atlantic hurricane, is shaped by basic principles and natural laws, why are things so unpredictable? Why do we perceive chaos and allow the thoughtless whims of butterflies into our equations? The answer takes us to the very heart of chaos theory.

Imagine a simple case of cause and effect: a leopard catches an impala; the impala dies. Admirers of the impala may cringe at the thought of the horribly violent death of this graceful antelope, but such is the nature of life in the wild. With a greater number of hungry leopards stalking the land, more impalas will meet their gruesome deaths. Fine, a simple principle. A mathematician would consider it to be a simple *linear* relationship—one could draw a straight line on a graph to represent the proportionate increase in leopards and decrease among impalas.

The equation of life, however, is not so simple and linear. Without enough food, in the form of impalas, the population of leopards may suffer. In nature, unlike capitalistic economic systems, the increase in demand does not ultimately result in an increased production of supply. Quite the opposite: increased demand almost always results in a decreased supply. So, based solely upon the simple principles espoused so far, the leopard population may decline due to lack of food as they deplete their food source of impalas.

What goes around comes around, or so it is said. And so it goes with the leopard and the impala. Fewer cats mean less threat to the impalas. Impala populations then rebound and fill the African landscape once again. This serves to restock the meat counter for the cats; they thrive upon the feast, and the impalas decline one more time. On simple principles, such a cycle of rise and fall can continue indefinitely. This simple yet ubiquitous ecological principle was noted by Alfred J. Lotka, a physical chemist, and his Italian counterpart, Vito Volterra, who formulated a mathematical model of the fluctuating relationships between predator and prey.

The laws of supply and demand, as envisioned in a simple Lotka-Volterra model, lead to a seeming balance of nature, albeit a cyclic balance. These cycles of population fluctuations do not follow a straight line but go up and down, depending on the densities of each population. They are *nonlinear*, as are chaotic systems. But if we add just one more principle, the picture becomes a bit more complex and the balance becomes a touch more chaotic.

A second principle could be this: a lion (rather than a leopard) catches an impala; the impala dies. And now two similar principles, those of leopard and lion hunting, both affect the fate of the impala. This is just the beginning of the chaos that is about to ensue for all the parties concerned.

One could complicate this dynamic model of life further, and an avid ecologist certainly would, to build an unpredictable chaotic system. Leopards, at least in southern Africa today, seem to have a strong preference for impalas, despite their fairly catholic dietary tastes. Lions, who ravenously consume everything from insects to buffalo, tend to opt for larger prey such as giraffes. Although this complicates our simple model of impala supply and demand, given the range of other options available, the impala situation would tend to affect the leopard more than the lion. The predictability of the system thus becomes considerably more complex and nonlinear, with population densities of each animal being dependent on the densities of all others.

Mathematical models of similar ecological relations have been constructed, and equations run for generation after generation in the simple mind of a computer. What happens? Well, it depends. Even a simple Lotka-Volterra model, as mimicked on a computer, can have many consequences. One can totally eliminate the element of chance with a strictly unwavering equation for the number of leopards, lions, and impalas, and still a plethora of scenarios may unfold—depending upon the *initial* numbers of each kind of animal. Throughout the generations, general oscillating trends may develop, and the system could go on forever. Alternatively, as we learn from both computers and real life, eventually the system may collapse, with one or more species going extinct. Such is the nature of chaos: unpredictability based upon the nature of initial conditions.

The true marvel of chaos, however, is that it can lead to ordered ecological systems and complex beings such as ourselves. Not just a few but thousands upon thousands of rigid ecological and evolutionary principles, perhaps complicated by chance and coincidental events, result in the balance of nature, or at least what we perceive to be a fairly stable situation. A chaotic system converges on a pattern known in mathematics as an *attractor,* and it can stay in that pattern for some time. Chaos then becomes an irony in and of itself, as the resultant system appears to be quite ordered. But the "balance of nature," if one wants to call it that, is a dynamic and nonlinear balance that can be frighteningly tenuous. We cannot predict the future behavior of the mathematical attractor, and it remains to be seen if we can predict what will happen next in nature.

HUMAN EVOLUTION has been the product of many forces that together made us neither inevitable nor probable. The links of the human evolutionary chain were riddled with chance, coincidence, and chaos, and we cannot fit the links together without a full appreciation of these factors. We can explore the initial conditions of our evolutionary past through the fossil record, and we can test for principles of what made us human. We may learn how the capricious forces of nature acted *constructively* in driving our evolution, and how they played integral roles in the Darwinian forces of natural selection. Science can find the mischievous factors, dissect their component parts, and dare to make predictions.

2

Between a Rock and a Hard Place

THE ROCK under my feet looks like soft, gray elephant skin, but for millions of years it has withstood the forces of time. The southern African winds and rains have weathered the surface, gradually taking it apart grain by grain. Over many epochs the occasional claw of an animal, and more recently the soles of human feet, have worn the rock here and there as well. The roots of the plants that grow on the hillside have cut into it and scarred it forever. But the rock has yielded little; it remains solid and hard.

Undaunted, we start the power drill. Pushing down on it with all our weight, and taking turns to conserve our strength, we try to bore a hole into the unrelenting gray rock, starting perilously close to a precipitous edge of the ancient exposure (fig. 2.1). Over the din our imaginations conjure up visions of precious discoveries. After about twenty minutes, the drill has progressed less than half a meter. That is far enough.

Into the hole are placed the steel "feathers" that guide a steel wedge. Once all is neatly fitted, we drive in the wedge with a sledge-hammer, again taking turns to elicit as much strength from our bodies as we can. Ten hits, twenty hits, sometimes more, and the resilient rock begins to crack. Hit again, and the crack widens. With some relief,

Figure 2.1 • Drilling through the dolomitic cave roof at the Makapansgat Limeworks, South Africa.

we watch as a chunk of rock breaks away, getting us just that much closer to the treasures that lie below.

Our efforts take us slowly through the roof of an ancient cave. Piece by piece we remove the hard gray roof and finally reach our destination—the reddish fill of the cave that accumulated millions of years ago. And what we find there is more rock. Through the ages the cave sediments have also turned to stone.

As we pull on the cord to start the drill's engine, our anxiety mounts. The red rock is somewhat softer and easier to drill than the roof but still takes its toll on our tiring muscles, especially with the hot African sun roasting our skin. As the sweat rolls over our brows and seeps into our protective goggles, it stings our eyes. At times like these we struggle to remember that paleoanthropology is thought of as a glamorous profession.

Ten minutes, and the hole is ready. Once again we carefully place the feathers and the wedge, full of anticipation. With the sledge-hammer gripped firmly in hand, the pounding begins. Our eyes widen as the crack begins to form in the rock. Fingers of light reach down through the crack, illuminating a glimpse of our past, like a candle in long-forgotten catacombs. Finally we release a sizable chunk of red rock from the cave deposit, get on our knees, and lift it from the edge with the strength we have left.

What will we find this time? All eyes inspect the rock slowly, carefully, and there is the small glimmer of enamel. The tooth of a prehistoric baboon, left behind in the cave sediments some 3 million years earlier, becomes reacquainted with the light of the African sun. Just a tooth, but more precious than gold to us.

As our eyes scan down the red rock we released, they pause at a bit of white streaked across the surface. Close examination shows the unmistakable structure of bone. Most of the bone is still encased in rock, and we see just a small bit exposed on the surface. Perhaps it is a bone from the body of the baboon; we cannot tell. Alternatively, it could be a bone from something closer to us. It could be a remnant of one of our own ancestors.

The teeth and bones we excavate are the product of ancient events. Long ago some hapless animals met their demise and were dragged into the dark reaches of a cave. Their bones eventually became encased in sandy rock; they remained hidden within the deposits until today. But a full view of the fossil remains will have to wait. In as little as one week, or maybe as long as a few months, they will be teased from the rock in our laboratory. Meanwhile our imaginations soar. We have exposed fossilized remnants of life from a key time in early human evolution.

Digging through rock, although frustrating, is the easy part. The true challenge, which fills us already with eager anticipation, will be to make sense of the teeth and bones we extract from the rock. Every fragment tells a story of former life and death. Together, all the fossilized animal parts from this cave and others nearby will fit into a jigsaw from which we shall someday envision past lifeways in the southern lands of Africa. Across the continent, more caves as well as ancient lake bottoms and river beds hold further clues that will help us reconstruct nature's creative processes. But for now the puzzle remains an incomplete jumble of pieces, difficult to interpret. We have no paleontological equivalent of the archaeologists' Rosetta stone to guide our interpretations;[1] we have only our imaginations and the scientific method.

Sitting at one of these ancient fossil sites, in the remote reaches of the country, we can easily slip our minds into the wild lives of the extinct animals we excavate: a chaotic menagerie replete with saber-toothed cats prowling the forest in search of our timid and scared ancestors. But much has changed since the time when these animals, like all others, struggled to survive. Even out in the field, with nothing in sight but African countryside, the contrast between life then and now becomes quite stark as a jumbo jet streaks across the sky. The jet is carrying one of the more peculiar products of evolution, descendants of the apelike prehumans who battled against long odds for existence when the caves of South Africa filled with sand and bones so long ago. What a difference just a few million years can make since the world first witnessed this odd, upright creature.

The Hard Place

The study of human origins has been around probably as long as humans have been sentient beings with language and imagination. It is part of our very nature to wonder about our past. But the science of human evolution is relatively new. Long after physics and astronomy had come to grips with many of the basic processes of the physical universe, biology was born as a distinct science. Thus, although Galileo

Galilei, the great Italian physicist and astronomer, was sentenced by the Inquisition for setting modern physical science on its current course in the 1630s, biology had to wait for more than a century to get into full operation as a rigorous study of the living world. Surely aspects of biology were known, but anatomy, for example, was part of the medical arts, and the natural history of animals was still part of natural theology.

It was in 1758 that Linnaeus gave us our system of naming biological species, such as the name *Homo sapiens* graciously granted to ourselves. Biology, however, was not given its name until sometime around 1800, thanks to the French scientist Lamarck, who could be considered the first biologist for having coined the term *biology*.[2] Lamarck also posited a theory of biological evolution. He was not the first evolutionist, but the predilections of a Greek philosopher named Anaximander, who sometime after 600 B.C. postulated that men evolved from fish, had long since been forgotten or ignored. Although Lamarck got the evolutionary mechanism wrong, he was responsible for getting evolutionary thought under way in the scientific world.

The processes and mechanisms of biological evolution, as we understand them today, started to become apparent only in the mid–eighteenth century, when Charles Darwin and Alfred Russel Wallace announced the theory of evolution through natural selection. This theory forms the cornerstone of modern biological science, comprising principles no less verifiable than those of physics. But in Darwin's time evolutionary theory, and particularly the idea of human evolution, needed some supportive data, hard evidence, before its key principles could be accepted by the scientific community. Such data initially came from fossil fragments, scattered across the globe. Now the evidence for evolution is overwhelming, but endlessly fascinating and full of surprises.

It was only in the 1850s that fossil humans were first recognized. Thomas Henry Huxley then brought theories of the evolutionary process to the human doorstep. Armed with the first known fossils of primitive humans, from the Neander valley of Germany, Huxley inexorably linked us to the continuous chain of life.[3] This was more than two centuries after Galileo linked the physical laws of the earth to those of the heavens. Finally paleoanthropology, the study of the evolution of fossil humans, was born. Somewhat later, in 1924, the caves of South Africa yielded the first glimpse of our most remote human ancestors, and their fossilized remains centered Africa on the map as the cradle of humankind.

Much has been learned since, even though only a small number of scientists have tried to tackle the questions of evolution and the origins of humanity. Our few scientific successes in paleoanthropology are fueled by fossil finds and driven by advances in technique, as well as by novel observations of all the life that exists on earth. Only now is the picture of human evolution, in its full natural context, beginning to come into clear focus.

On the other hand there is much more to be learned. Each answered question about our origins has spawned new, more detailed and challenging questions. What led our ancestors to stand up on two legs? What forces propelled the evolution of our ancestors' brains to unprecedented levels of size and complexity? Only with more digging and thinking can the vision of our past be constructed with more certainty. The past is indeed the greatest unexplored frontier.

As our young science matures, it also ripens for change. As we learn more and develop new methods of looking into the past, we erase some of our former knowledge even as we build upon it. This often happens in science, for the scientific method is a self-correcting process. Science itself evolves. Our explorations at the wild frontier of our evolutionary origins will bring new and perhaps startling insights. Old theories will crumble, much as once mighty species have gone extinct. New concepts and bold hypotheses will arise, to be tested against the rigors of the scientific method, not unlike the ways in which the surviving life forms of today have endured challenges to their very existence.

Some truths will remain, even though never quite escaping from their status as theory. It is still a theory that the earth rotates around its axis and revolves around a star called the sun. Such a theory, heretical in its day, is now fairly well established. It *could* still be wrong, but the chances of that diminish every day as the sun "rises" and "sets" in accordance with predictions of the theory. Likewise, in evolutionary science, there are established theories and those which are more open for serious debate, leaving room for varied interpretations of the evidence. Indeed, paleoanthropologists have a myriad of ideas, and a well-earned reputation for being loath to admit that they agree on anything. In this book we'll explore some of our latest heretical notions concerning the origins and evolution of our species, and continue the debates on how we evolved to our unique place in nature.

Nevertheless, there are solid cornerstones of principle on which all reasonable scientists can agree. As biologists, as evolutionists, and *even* as paleoanthropologists, we hold the following truths to be self-evident: evolution is a fact, and evolution is a theory.

The Fact of Evolution

Evolution is a fact.[4] By definition, evolution simply means change through time, and change is observable. Long before a chap named Charles Darwin came along, evolutionary change was readily apparent to laypersons and scientists alike. Domestic animals provided prime examples of populations that had changed—populations that evolved. Those same examples stand before us today. Greyhound dogs, Jersey cows, and Siamese cats are all animals of a particular type that did not exist before humans came along to cultivate them. Their respective populations then changed, largely due to the consequences of human meddling in nature's reproductive affairs. You will not find the bones and teeth of any of them in the fossil record, at least not in the particular forms they have taken today. Their changes are largely historical and are documented to some degree. So evolution in its strictest sense has occurred.

For biological organisms it is change in the fundamental genetic composition of a population that really defines evolution. At its most basic level, a population can be characterized by genes. Genes are the ultimate units of heredity, passing information for bodily construction from generation to generation. The relative frequencies of certain gene types have changed through time, at the expense of other types. A population of greyhounds is derived from a different set of genetic variants than a population of boxers; that is why their bones and teeth and coats are different from birth to adulthood. We humans have a habit of continuously breeding novel and presumably useful life forms for our own uses. Thus we can watch evolution happen before our very eyes as the composition of genes within a species changes.

A striking example of evolution as unequivocal fact is the change in genetic frequencies within bacterial populations, especially disease-causing types. Bacteria evolve, and they do so rather rapidly. They exist in phenomenally large populations, living by the millions in each human body and by uncountable numbers across the planet. Many types can be killed with the use of antibiotics, whereas a few others survive our medicinal onslaught. Within the populations of bacteria, there are more floating around now that are resistant to medical antibiotics than there were before we started using such medicines on ourselves (and more recently on our domesticated livestock). We humans introduced selective breeding again, although in this case we did it somewhat less consciously, oddly selecting for those bacteria which are most likely to continue to infect us. The result has been bacterial evolution, pure and simple. Fortunately there appear to be limits, or at least stumbling blocks, to the evolution of bacteria, and our medicines still do us some good, for now. But the bacteria are still evolving within each of us every moment.

The fact of evolution can be observed over and over again in many life forms, ranging from bacteria to cockroaches to pigeons to dogs. Even the modern human population seems to be changing still, evolving day by day in subtle ways as some are fit enough to survive and reproduce while others do not contribute their genes to future generations. Our evolution is not complete, and cannot be complete. All life forms continue to change, at least at the level of gene frequencies. That is a fact.

The Theory of Evolution

Evolution is a theory as well as a fact. The theory of evolution is Charles Darwin's notion that species owe their origins to evolution through natural selection. A greyhound and a boxer are both dogs; they are members of the same species, *Canis familiaris*. Their populations are different in their morphology and their behavior, largely due to conscious human selection, but they are the same species. A boxer and a greyhound can mate and produce offspring, a perfectly viable and happy puppy representing *Canis familiaris*. But the evolution of a new *species* is not necessarily observable, at least not in the short span of endeavors in evolutionary science. Could a pet dog, a wolf, and a jackal, all different species, have the same origin? Could a wolf and a jackal arise in nature, without a conscious helping hand? Could humans and chimpanzees have performed the same evolutionary trick of divergent speciation from a common ancestor?

These questions fall into the realm of theory, for such events cannot be directly observed. This theory of the evolutionary origin of species, like any theory, cannot be absolutely proved. Theories, from the "hard" sciences such as physics to the "soft" sciences such as sociology, can never attain absolute proof, which exists only in pure mathematics. And even there it can be argued that the conclusions are somewhat suspect.[5] Certainly in the difficult science of evolutionary biology, proof is rather elusive.[6]

Yet we do have good ways of testing the predictions of Darwin's theory. The best way is to use the bones and teeth in the rocks we excavate as evidence. Quite simply, Darwinian theory predicts that as we go back in time through the fossil record, life forms should look more and more different from those existing today. Indeed they do. Fossils can be deceptive, but they do not lie; and they tell a fascinating story of evolution. In this book we shall take a close look at some of this fossil evidence and delve into its very nature. But for now I can tell you that all our rock hammers and power drills have not even dented the rock-solid structure of evolutionary theory; every fossil we have found supports it. Evolutionary links to the past are missing only in detail.

Other tests of evolutionary theory exist as well, including close observations of the very genes that characterize evolving populations. Here we get into higher technology and sophisticated techniques, ranging from identifying the chemicals that make up the genes to computerized simulations of changes in those genes through time. We'll experiment with some of this evidence as well in subsequent chapters, including some novel and enlightening computer simulations. Once again we shall find that the nature of genes, the very code of life, supports the notion of Darwinian evolution.

Darwin's theory goes further, of course, than merely predicting change. It was in 1858 that Charles Darwin, along with a lesser-known scholar named Alfred Russel Wallace, announced the theory of the origin of species *through natural selection*—a powerful mechanism that is constantly working in nature. With domesticated animals, as mentioned above, we see the evolution of dog and cattle populations by means of artificial selection. Darwin and Wallace proposed that a similar mechanism produces new species in nature, without the guiding hands of humans. Darwin himself dubbed the process "natural selection."

The logic of natural selection is terribly simple, and thus tremendously powerful. When a being is born, it has a chance to live and reproduce. Every being is unique, and thus some beings may have a greater chance than others to carry on. Natural selection begins. Perhaps they are better adapted to their surroundings, or maybe they are adaptable enough to thrive in a wide range of environments. Either way, as long as they contribute their essence to the next generation, through the genes that code for aspects of their success, the process of natural selection continues. The most successful variants survive. Eventually there is a chance that one group of variants will become a new species, no longer breeding with other populations. Those other groups may become extinct, or they may produce different variants that survive for other reasons, evolving into other new species. In natural selection, nothing succeeds like success.

Think of natural selection the next time you get an infection. Here I mean natural selection not of our own species but among the varieties of bacteria that cause a particular infection. As we now know, more and more types of those bacteria can survive in your body even when your doctor gives you penicillin or any other antibiotic. We did not consciously select them for proliferation the way we bred cows that produced more milk or dogs that won prizes for physique. But our hand was active nevertheless in their evolution, for we administered the antibiotics; nature then bred the winning bacteria. Nature is relentless.

One of the early deterrents to the acceptance of natural selection as a key principle in the origin of species was that Darwin could not fathom the *source* of the variants on which natural selection acts. He rightly noted that variability existed among living creatures, but its sources and limits remained elusive in his day. The confusing, disorderly rise of variations prompted one of Darwin's respected scientific colleagues to call natural selection the theory of "higgledy-piggledy."[7] Darwin and his contemporaries knew not of genes and chromosomes. Without knowledge of genes, they could not conceive of genetic mutations, the ultimate source of all biological variability.

Now we know that mutations of genes, largely at the whim of chance, produce the variations upon which natural selection acts. It is random genetic mutation that ultimately produces drug-resistant bacteria, high-yielding cows, and large human brains. Despite the tendency for most mutations to decrease the viability of an individual, occasionally a new gene results in a new feature that coincides with an individual's ability to survive. To survive *and reproduce,* that is. Through reproductive success, nature favors that individual and blindly promotes the new gene as it gets passed on through subsequent generations.

The theory that natural selection drives the evolutionary origin of species is well established. All our scientific tests, based on fossil data or genetic experiments, appear to verify the continuous process of natural selection. There is no doubt that Darwin and Wallace were on to a very important phenomenon. Evolution through natural selection is indeed now entrenched in the foundation of contemporary biology, and there it will probably stay.

Nevertheless we can, with all due respect to Darwin and the many biologists who followed, ask some challenging questions of natural selection. We can chip away at the theoretical foundation and probe its depths. That's what science is all about! Shaking up the world of science can be as much fun as discovering fossils, even if both are sometimes difficult to do.

First of all, modern evolutionists have a small problem with the term *natural selection*. It is ingrained in our lexicon, so we accept the term and use it liberally. But it is a touch misleading. *Selection* has a teleological connotation of design or plan, as if a human hand (or some other consciously driven hand) is guiding the process. But in evolution there are no designs; there are only consequences. Natural selection merely ensures nothing more than the coincidence of the survival of the survivors; if a being can carry on in a given environment long enough to reproduce and make more like itself, then so be it. There is no external selecting entity, just an intrinsic force with no particular direction beyond survival and reproduction. Darwin himself eventually realized that *natural selection* was a misleading term and suggested instead *natural preservation*.[8] But the die was cast, and the name stuck.

There can be no reasonable quarrel with Darwin that natural selection, or whatever term one may choose to use, is a powerful mechanism. The idea will pervade the rest of this book. Evolution cannot proceed fruitfully without natural selection. But there is more. Thomas Henry Huxley, one of Darwin's contemporaries (and one of my scientific heroes) suspected that natural selection alone was not adequate to explain life on earth.[9] More principles were involved. Wrote Huxley:

How far "natural selection" suffices for the production of species remains to be seen. Few can doubt that, if not the whole cause, it is a very important factor in that operation; and that it must play a great part in the sorting out of varieties into those which are transitory and those which are permanent. But the causes and conditions of variation have yet to be thoroughly explored; and the importance of natural selection will not be impaired, even if further inquiries should prove that variability is definite, and is determined in certain directions rather than others, by conditions inherent in that which varies.[10]

A partial retort to Huxley's caution about the "causes and conditions of variation" was provided earlier this century. Beginning in the 1920s, genetics and evolutionary theory merged into a scientific synthesis that guides our understanding of life today. Genes hold the blueprint for development of an organism and are responsible for much variation. With the discovery of their role, many were satisfied that Huxley's reservation about natural selection and queries about the sources of variation had been answered. But one may wonder if Huxley had envisioned even more. Some evolutionary scientists, myself included, suspect that the evolutionary synthesis of the 1920s is not complete and must venture beyond the genetic blueprint.

Thomas Huxley had nothing against Darwin's notion of the evolutionary origin of species through natural selection. Indeed, Huxley was known as "Darwin's bulldog," for while Darwin sat quietly at his home in Down, England, meticulously studying barnacles and other peculiar life forms, Huxley was out defending the cause of evolution against onslaughts from scientists and clergy alike. Huxley himself coined the term *Darwinism*, so his query on the *relative* importance of natural selection could hardly preclude him from being the first Darwinist.

The power of Darwin's argument could not be muted for long, especially with the likes of Thomas Huxley cleverly explaining it from every conceivable angle. When viewed in the light of evolutionary theory, fossil successions began to make enormous sense. But Huxley, good scientist that he was, questioned everything. Skepticism was seen by Huxley to be a "duty" of every scientist.[11] Natural selection, despite its explanatory power, is not exempt from the most rigorous scrutiny. Today we must question it still, despite our comfort with the concept.

What intrigues me most is the part of the Darwinian dogma that we most often overlook: despite the power of natural selection, it sometimes fails. After all, species do not always adapt but often go extinct. And even surviving species, including our own, are less than ideal. They possess many features that are poorly adapted—despite the persistence of natural selection. This is part of what Huxley may have meant by stating that "variability is definite": it has intrinsic limits and is far from perfect. Moreover, in some cases natural selection may not have just been inadequate; it may actually have caused extinction or maladaptation. If all evolutionary change is due to natural selection, then nature's selective criteria are very strange indeed.

Perhaps natural selection, albeit important, is not as important as our research has led us to believe. An analogy may help to elucidate the source of our scientific bias. Have you ever noticed that leaders of corporations, institutes, or even countries frequently take credit for everything that goes right but pass the blame for things that do not work out as planned? We tend to cultivate some sort of mystical connection between a person in power and the positive things that happen, even if no true link ever existed. I suspect that natural selection is the proverbial leader of evolutionary theory. It gets much more credit than it deserves. Most biologists tend to attribute our every adaptation, indeed almost every nuance of our morphology and physiology, to the action of natural selection. Natural selection has become an assumption rather than a theory.

Huxley warned us against this: "it is the customary fate of new truths to begin as heresies and to end as superstitions."[12] Thus, when Huxley wrote those words on the twenty-first anniversary of the publication of *Origin of Species*, natural selection had already taken on mystical powers and mythical proportions in the eyes of those who accepted evolution as truth. Huxley continued his dire admonition, pointing out that "irrationally held truths may be more harmful than reasoned errors." Darwin's bulldog then challenged the scientific community to challenge natural selection. It is, after all, a testable theory.

In the true spirit of Huxley, I shall argue throughout this book that it is impossible for natural selection to work quite so well as some suspect. The primary reason is simple: the limits are imposed by the very nature of chaos. Indeed, as we shall see, the chaotic nature of natural selection goes deeper than the nonlinear ecological systems in which animals are evolving, to the very heart of genetic evolution and morphological development. Some cherished notions of the causes of evolution, and of human evolution in particular, must then be laid to rest for their overreliance on natural selection.

Natural selection sometimes fails, and fails for some very interesting reasons. These failures comprise a leitmotiv of this book. But we must also search for other modes and explanations of evolution. Huxley was not able to offer an alternative to natural selection, aside from the enigmatic "conditions inherent in that which varies." But with the hindsight of scientific progress on evolution we might be able to add substance to his arcane phraseology. Yes, natural selection was important for the origins of species as we know them today, but there is more.

We now know that the origins of variation, in the form of genetic mutations, are due to chance. We also know that those few mutations which add survival value must coincide with a suitable environment. Part of the success of evolution is thus mere coincidence. The evolutionary process that then ensues is unpredictable, and in modern parlance could be considered to be chaotic. Nature thus "selects" more than just a survivor. It blindly determines contingencies for the future path of evolution—it sets the initial conditions.

Even under the relentless guidance of natural selection, the more whimsical and undirected forces of evolution can then dictate the course of things to come. The unlikely becomes what is likely to happen, and the seemingly probable becomes less so. The infallible drive toward greater fitness takes a strange turn, and natural selection seems to falter. A butterfly flaps its wings, a startled impala escapes a stalking leopard, and the course of evolution changes forever. Surprising things happen due to the niggling, mischievous, capricious nature of three very important components of evolution: chance, coincidence, and chaos.

You and I are among their products.

EVOLUTION MAY BE a fact, but detailed facts of our human evolutionary history and prehistory are sparse. Much of the information we need to chart the full course lies buried in the fossil deposits of eastern and southern Africa, the cradle of humankind. Much more has been lost forever or remains encased in ancient rock we have yet to find, giving us a woefully incomplete fossil record. Thus our first task is to excavate rock, get some basic data, and generate ideas (chapter 3). It will become evident that to interpret the fossils correctly and test our ideas, we must be cognizant of the role of chance at three levels: chance in the mechanisms of fossilization, chance in discovery, and chance in the evolutionary process.

Throughout this book, we shall try to bridge the gap between fossil-bearing rock and the hard place of theory. The information we find in fossilized bones not only provides a sketch of our past; it also allows us to test Darwin's and others' theories of the evolutionary process. One such theory, indeed the one that dominates many scientists' view of evolution, is that climatic change drove evolution and was responsible for the origin of humans. In chapter 4 we test whether the correlation between climatic change and human evolution is an important consideration or a mere coincidence. Toward that end we shall employ a creative paleontological approximation of the Rosetta stone—computer simulations that bring the ancient bones to life. Then, in chapter 5, we'll look in greater detail at some other theories that bear on human origins. It should become apparent that what is lacking from current ideas is a full appreciation of chance, coincidence, and chaos.

The next step of our analysis, in chapter 6, will be to build a conceptual model of the evolutionary process from the ground up (rather than from the skies downward). We'll start with the genes and find that their evolution not only includes chance, coincidence, and chaos but depends on them. Yet genes are only part of Huxley's "conditions inherent in that which varies"; in chapter 7 we'll build human bodies from their genes and see how bodily development itself is a chaotic system with intrinsic limits. However, the chaos also provides opportunities for evolutionary novelties.

The workings of chance, coincidence, and chaos, together with Darwinian natural selection, lead to a conclusion that evolution is self-propelled, or autocatalytic (as are many chaotic systems). This notion is fully explored in chapter 8.

Finally, in chapter 9, we'll look at the implications of our evolutionary conjectures for understanding the human place in nature today, and we'll look to the future as well.

The story of our distant past, and the story of our young science, begins in Africa. Armed with a battery of excavation tools, we start at a remote African village where the rock turns to dust.

3

A Tale of Two Sites

DUST. Whenever I think of Buxton, I always think of dust. Kalahari winds blowing dust into my eyes. Lime dust caked into my hair. Dust settling in my beer. Dust in every imaginable place and a few unimaginable places as well.

Despite the omnipresence of dust, there *are* more pleasant and interesting features. Buxton is a small African village of congenial Tswana people in an area known as Taung, placed quite literally at the southeastern margin of the Kalahari desert. An escarpment demarcating the edge of the desert is the only feature other than the scraggly thorn trees to break the flat monotony of the landscape. Yet Taung has a stark, desolate beauty that has never failed to enchant even the least romantic visitors I have taken there.

Taung is close to nowhere. Almost untouched by the urban products of humankind, it holds nature and humanity in its raw form. In the evenings, the wind carries not only dust but voices and children's laughter from the village of Buxton below, up to the top of the escarpment where I sit, gently puffing my pipe and nursing my beer. One can listen as another generation of people are growing, ready and eager to take on the challenges of life at the edge of the desert.

There on my favorite perch, I sometimes tune out the sounds of my human friends and focus my attention on the local troop of chacma baboons as they settle on the cliff edge for the night. I have spent many evening hours transfixed by their antics. So much like us they sometimes seem, in their social nature: the young playing in the dimming moments of daylight, the adults keeping a watchful eye on their progeny. With a diminishing litany of sounds from their vocal repertoire, the baboons become quiet as the sun emits its last light over the Kalahari. As dusk turns to darkness, only the occasional screech of an owl or the distant whining of a jackal breaks the silence.

The Southern Cross, always the first constellation of stars to appear, then emerges in the night sky. First the brilliant pointer stars, Alpha Centauri and Hadar, and then the four stars of the Cross itself. Soon the sky fills with stars, and the brightness of the Milky Way begins to obscure the distinctiveness of the Southern Cross. Looking to the north, late at night, one can see another galaxy, separate from our own Milky Way—the Andromeda galaxy, glowing obscurely. As the apparition takes form, I wonder what life was like here on earth when that light was emitted by the stars of Andromeda, some 2.5 million years before it reached my eyes at Taung, above the village of Buxton.

With the brilliant stars, the barks and grunts of baboons, the crisp air of the desert night, one could go as far as to say that life in the village of Buxton is attractive. It has been especially attractive to an anthropologist like me, for it was in this area, long ago, that our prehuman ancestors struggled to lay the foundations of our existence today. Along with them lived the ancestral baboons and precursors of many other African creatures that determined a crucial part of the evolutionary course to come.

Nearly 3 million years ago the Taung area would have been very different indeed, unlike the dry, dusty land we know today. Life apparently flourished there. A strong spring at the edge of the escarpment would have fed diverse plants and animals. Antelope grazed on the savanna grasslands of the sprawling flat landscape below. Not one but two species of baboons climbed across the rocky slopes to quench thirsts born under the hot African sun. Leopards, lions, and perhaps saber-toothed cats would have found sufficient prey on which to feast, more often than not an unfortunate baboon. Hyenas, jackals, and scavengers from the skies would have been eager to consume the remains. The cats, having satiated their appetites and wary of losing their prey to scavenging, dragged the carcasses of their victims into the local caves. These cool, dark havens, hidden behind the water flowing over the escarpment edge, provided a nice place to hide a future snack.

Among the prehistoric animals seeking water, food, and shelter came some very novel and peculiar creatures. They walked precariously on two legs, balancing their modest brains almost directly above their shoulders. Although they shared habits with the baboons, eating the same roots and seeds and fruits, they were slow walkers and clumsy climbers. Exposed to the harsh elements and vulnerable to the abundant predators, with little recourse for escape or protection, their chances for survival seemed slim. Yet somehow these slow animals of marginal intelligence carried on, and we are living proof of their unlikely resilience. They were our ancestors.

One could not say that these early predecessors of the human race thrived. Their numbers seem to have been kept low in the Darwinian struggle for survival. Some such mishap undoubtedly befell a young child in these early days of Taung. He was only four or five years old but a bit precocious by modern standards, and ably using his two legs to walk around the rough terrain of the escarpment by himself. Having a childlike curiosity, so characteristic of human nature, he wandered away from the relative safety of his family. Perhaps a leopard could not resist this easy morsel of flesh; it was usually the young, the weak, or the old that served themselves as lunch to unbemused predators. Or the boy may simply have slipped while trying to get a drink, and haplessly drowned in the pool of cool water. We will probably never know.

Whatever circumstances led to the premature demise of this prehistoric child of Taung, his bodily remains floated in the water and washed back into a cavity within the rock. His family may have searched for him, to no avail. Eventually they would have sauntered off in search of food, unaware of the profound influence their lost child would have on twentieth-century science and unconcerned with such fates as well. Meanwhile the child's body sank into the sediment. The waters stirred sand into his decaying skull, and he lay still for ages, cheek to cheek with young baboons and other animals that had met a similar fate.

The inexorable arrow of time on earth drove many changes at Taung. Long after the days of our earliest ancestors and their forgotten child, the water at the edge of the escarpment deposited sediments that sealed the cave in which the child's remains had been entombed.

New caves formed, again capturing the most dramatic results of the struggle for existence in a world where beast consumed beast. But less and less did early humans, if we can call them that, find themselves the victims in nature's web.

Throughout the phases of environmental change to which the African continent was subjected, Taung saw the presence of humans. In times of cold and times of warmth, with fluctuations from the very wet to the very dry, women and children and men eked out an existence, ultimately fed by the perennial springs. Where there is water there is life, and though the quantities varied throughout the millennia, the most adaptable animals stuck it out at Taung while the more specifically adapted creatures came and went. Some went forever. Our ancestors, the descendants of the lost Taung child's family, were among the adaptable. First they developed crude stone tools, and eventually they captured fire, fashioned clothes, and constructed shelter. They were survivors. We owe our very existence to them.

Precious Things in the Earth

I lived in South Africa through a period of dramatic change, in the late 1980s and early 1990s. The country became the world's newest democracy, and native Africans were freed from the oppressive rule imposed by people originally from afar. A new way of life began for millions of human beings. But profound change has been part of the experience for humans and their ancestors in southern African over the past 3 million years. Evidence of these changes chanced to be recorded, sporadically and unsystematically, in the caves at Taung.

The secrets of the caves were first revealed in 1919 by a long series of fortuitous circumstances, oddly initiated by the unique human susceptibility to gold. Gold became part of the foundation, both literally and figuratively, of the city of Johannesburg. The drive for gold had repercussions. The area of Johannesburg may have been rich in ore of precious elements, but less precious matter was necessary to sustain human lives and economic productivity. For example—and this example is not arbitrarily chosen—lime was needed. Limestone, calcium carbonate, was initially used in the processing of gold. Lime was also necessary for cement to build the structures that house humans and their industries, and lime helped enrich the agricultural land that ultimately fed modern humans. Lime was abundant at Buxton, Taung.

A coincidence of local geology and human need suddenly thrust Taung to importance in 1916, shortly after gold was discovered in Johannesburg. Exposed on top of the escarpment at Buxton, and underlying much of the Kalahari desert, is a bedrock foundation of dolomite. This dolomite was formed over 2 billion years ago at the bottom of a sea, enriched for millions of years with the lime from marine shells accumulating there. The exact processes leading to its formation are still a bit of a mystery, but eventually the earth lifted the dolomitic rock from the seabed to form part of the continent of Africa.

By its geologic nature, dolomite contains layered cracks through which rainwater can percolate from the surface and accumulate below as groundwater. During this journey water dissolves the dolomitic lime and can form tremendous caves. As fossil hunters later discovered, these caves underlie much of the South African landscape and hold many scientific treasures. But at Taung, the groundwater in the dolomite had a distinctive destination.

Sometime in the distant past, a shift of the geological crust on the African continent caused a fracture along a line of weakness. Rock was lifted along one side of this fault and created the escarpment at Buxton. Perhaps by meteorological chance, perhaps by default, that escarpment now forms the southeastern margin of the Kalahari desert. On top of the escarpment is lime-rich dolomite. Within the bricklike network of dolomitic blocks, lakes of groundwater fill the seams and ultimately emerge as the tireless springs that feed the land and life below.

The water exuding from the rock carries dissolved lime, a calcium bicarbonate solution. Indeed one can taste its distinctive tanged flavor with every drink and feel the residue it leaves on the skin after a swim in the spring-fed pools. As this water spreads out over the escarpment, its flow reduces to a trickle at the periphery of the river. There moss and algae thrive, concentrating the lime from the water. In drier periods the sun bakes the moss and algae, leaving behind plant skeletons constructed of soft, white lime—calcium carbonate. Season after season, year after year, such skeletons accumulate on top of each other. Like heavily starched white sheets draped over an irregular bed of rock, layer upon layer of lime grows into thick deposits.

The layered lime solidifies as a rock called tufa. Some speak of "tufa flows," for the resulting masses give the appearance of having flowed like lava, as if spread like white icing on a cake. But tufa is really an accretion of layers, and each layer captures a momentary construct of life. Henry David Thoreau could have been describing tufa when he wrote: "The earth is not a mere fragment of dead history, stratum upon stratum like the leaves of a book, to be studied by geologists and antiquaries chiefly, but living poetry like the leaves of a tree, which precede flowers and fruit,—not a fossil earth, but a living earth; compared with whose great central life all animal and vegetable life is merely parasitic."[1]

Such a romantic view of the massive tufa accretions at Buxton could not dissuade progress—not when there was a country to build. Geologists came and noted dispassionately what millions of years of living earth had built for the taking: soft, pure lime. And in 1916 quarriers arrived and began deconstructing the noble achievements of moss, algae, water, and sun. It was not until 1919 that antiquaries also took interest in the contents of the tufa.

For a few years the tufa at the Buxton Limeworks was a quarriers' delight. With each blast of dynamite tons of limestone crumbled to the ground, where it could easily be collected and taken to the kilns for processing. But in 1919 the quarriers encountered what could have been called the red menace: within the tufa were pockets of red sandstone. Although the sandstone was cemented by lime, it did not contain enough calcium carbonate for economically viable processing. Some sandstone pockets were small, but some were large and particularly menacing to the economic progress of the quarry. Thus some sections of the quarry wall were avoided altogether and left standing. Other spots of red sandstone were blasted out and discarded in massive dumps, except for a few choice pieces, kept as curiosities, which encased fossil bones.

These bones had arrived by peculiar means. When the tufa formed it did not always do so in neat, uniform layers. As accretions built out over the edge of the escarpment, they sometimes left behind open cavities (fig. 3.1). Much in the way that icicles or cave stalactites form as hanging cones or sheets, tufa curtains could build out over an edge and hang in front of cool, shaded pockets. Sometimes these curtains formed the roof and walls of substantial caves. The winds and the backwash of water carried the red sands of the Kalahari into these caves. Animals, attracted by the shade and coolness provided within, used the caves for shelters or feeding lairs. With time, the caves filled with red sands and bones until the continuous building of tufa closed the curtain and sealed the cave. Water percolating through the porous tufa

Figure 3.1 • A tufa cave formation near the Taung fossil site. The river reaches about six meters back under the tufa.

brought lime to cement the sand and fossilize the bone. A quarrier's red menace, and a paleontologist's delight, had been created.

Most of the fossil bones the quarriers occasionally encountered were the remains of extinct baboons. Beautifully preserved skulls revealed a wealth of anatomical detail by which these early creatures could be distinguished from their modern descendants. This window on the past stirred the imagination of a young science student named Josephine Salmons.

The involvement of Josephine Salmons in the discoveries at the Buxton Limeworks was full of twists of fate, at least as far as her professor, Raymond Dart, was concerned. Salmons was registered for a bachelor of science degree at the University of the Witwatersrand Medical School in Johannesburg. She and her fellow students had been urged by Professor Dart to collect fossils during their July vacation. The most fascinating fossil that Salmons encountered, however, was not imbedded in some remote rock deposit but was displayed on the mantle over the fireplace at the home of a family friend, Pat Izod.

E. G. Izod, Pat's father, was a director of the Northern Lime Company, the managing company of the Buxton Limeworks. While visiting the quarry in early 1924, he was shown an interesting fossilized monkey skull, which he kept as a curio. Fortunately it later caught the eye of Josephine Salmons, and she correctly assumed that it would please her dynamic professor.

The fossil skull aroused Dart's insatiable curiosity, for ancient remains of primates (monkeys, apes, and humans) were then rare in Africa, particularly in southern Africa. The skull was clearly that of a baboon yet recognizably different from the baboons of today, even to an untrained eye. Dart asked that other such interesting skulls from Taung be sent to him. Fortuitously (and this story is full of fortuitous events), a geologist from the university was traveling to Taung in November, as a consultant to the Northern Lime Company. This geologist, Robert Young, was to be yet another link in a complicated and coincidental chain of discovery.

Upon his arrival at Buxton, Young visited the office of A. E. Spiers, the quarry manager, to relay Dart's request for cooperation in the collection of interesting fossils. In Spiers's office were a number of skulls collected by a crusty old miner, Mr. De Bruyn. One particularly large piece, from a recently exposed cave fill, served as a paperweight on Spiers's desk. Professor Young gratefully packed some of the skulls into a box for Professor Dart and, with characteristic thoroughness, relieved a reluctant Spiers of his paperweight.

The fossils were crated and sent by train to Raymond Dart. An acknowledged expert on the anatomy of the brain and a self-professed expert on many other things, Dart was well prepared to assess the contents of this particular crate. On top of the pile of rocks was Spiers's large paperweight; Dart immediately recognized it as an endocast of a skull. It had formed when the skull of some creature, larger than a baboon, had filled with sediment. The skull bones broke away in the blast that released the rock from the quarry, leaving a rock endocast— a perfect cast of the impressions on the inside of the skull. Mammalian skulls are impressed with the convolutions of the brain, so Dart was able to read the features of the brain on this particular endocast. The story the endocast told was to change our view of human evolution forever.

Dart excitedly fished through the crate and found a piece of rock in which he could see the back of some facial bones, as if viewed from inside the skull. This rock fit the endocast perfectly. So, as the legend goes, Dart sharpened his wife's knitting needles and proceeded to

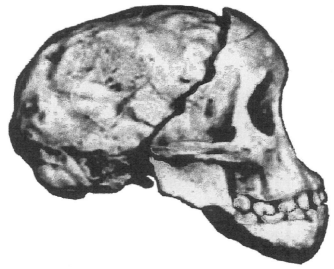

Figure 3.2 • The Taung hominid skull.

expose the front of the face by carefully chipping away at the encasing rock matrix, until he revealed the face of the long-forgotten child of Taung (fig. 3.2).

The clues revealed by the beautifully preserved face and endo-cast of the Taung child were overwhelming, at least to Raymond Dart. The face had no snout like a baboon's but gently curved out around the jaws. The first permanent molars were emerging in the back of the mouth, as they would in a six-year-old human child, but this was no modern human child.

The brain, as revealed by the endocast, was certainly larger than that of even an adult baboon, but it was scarcely larger than the brain of a modern chimpanzee of the same developmental age. This con-fusing array of features started to make sense when Dart noted the position of the foramen magnum, the big hole at the base of the skull where the spinal cord merges with the brain. It was much farther for-ward on the skull base than anybody would have suspected.

In a quadrupedal animal, one that walks on all four limbs, the foramen magnum is at the back of the skull. The horizontally placed spinal cord continues straight forward from the vertebral canal through the hole to the brain. This placement would not work in humans, a bipedal animal that walks on two legs and carries itself with an upright posture. The human spinal cord runs vertically up to the foramen mag-num to meet the brain near the center of the base of the skull. Thus the big hole is positioned farther forward than that of a quadrupedal ani-mal. In this respect the Taung child was much more humanlike than any quadrupedal mammal of its time. Indeed, it was much more humanlike than today's knuckle-walking apes.

Dart was possessed by the significance of the Taung child's anat-omy. No one had ever seen such a small-brained upright animal in the fossil record. Few suspected that such a creature would ever have been found in Africa. Dart wrote a careful description of the fossil for the British scientific journal *Nature*, in which he named a new genus and species for this unique creature: *Australopithecus africanus*.[2] *Austral* refers to southern, *pithecus* means ape; so Dart named the Taung child the southern ape of Africa. But despite the name, he really meant that it was the southern ape-man of Africa, a representative of our earliest ancestors.

Major shifts in scientific thinking often take hold slowly. Early support for Darwin was certainly not overwhelming, and for that mat-ter Galileo and even Newton never quite convinced all their contem-poraries. Raymond Dart was in for a rough ride as well. For those who accepted evolution back in 1925, it was the expanded brain that made us human, not upright posture. And humans first evolved in Europe or Asia, certainly not at the bottom of Africa. Dart could not have found an African ape-man; clearly he had been confused by the undeveloped anatomy of this aberrant African ape, with a skull so different from what one would expect in an adult.

The pages of *Nature* soon filled with comment and speculation about Dart's "ape." Most correspondents were rightly cautious or politely dismissive, except Arthur Keith, an aspiring leader of British anthropology. Keith was conciliatory in stating that "Prof. Dart has made a discovery of great importance, and the last thing I want to do is detract from it," yet he called Dart's claim for the position of *Australopithecus africanus* in human ancestry "preposterous." Dri-ving his point home with an assumed very recent age of the fossil, Keith wrote: "A genealogist would make an identical mistake were he to claim a modern Sussex peasant as the ancestor of William the Conqueror."[3]

Professor Grafton Elliot Smith, however, noted that the scientific community should not be surprised by Dart's discovery—not if they "know their Charles Darwin."[4] Elliot Smith, an old friend and mentor of Dart who had first suggested to Dart that he apply for the chair of anatomy at the University of the Witwatersrand, was referring to a deduction about human origins Darwin had expressed in *The Descent of Man* in 1871:

> In each great region of the world the living mammals are closely related to the extinct species of the same region. It is therefore probable that Africa was formerly inhabited by extinct apes closely allied to the gorilla and chimpanzee; and as these two species are now man's nearest allies, it is somewhat more probable that our early progenitors lived on the African continent than elsewhere.[5]

In the years to come Dart and Darwin would be vindicated, largely due to the efforts of a feisty old physician-cum-paleontologist named Robert Broom. Broom championed Dart's hypothesis of an African origin for the "ape-men" that evolved into humans and was as pugnacious in his defense of Dart as Huxley had been of Darwinian thought. Broom's subsequent discoveries of adult *Australopithecus africanus* fossils at the South African cave site of Sterkfontein, and later fossil finds across Africa, proved that Dart's reading of the Taung child's anatomy was as brilliant as it was prescient. There is now little if any doubt that humans originated from small-brained, large-faced, bipedal australopithecines. Dart was the first to establish that our ancestors jumped onto the road of human evolution feet first. They were not using their heads, at least not much.

In retrospect, Dart should not have been surprised by the discovery of a biped with a relatively small brain—not if he knew his Lamarck. Jean-Baptiste de Lamarck, a pre-Darwinian French evolutionist, had predicted in the early nineteenth century that bipedalism preceded the expansion of the brain.[6] It was quite a prescient idea as well, given the lack of any fossil evidence, but one vindicated by Dart's marvelous fossil child.

One can reflect on the fortuitous and serendipitous events that led to the discovery of the Taung child, the first of the African "missing links." A host of characters—Josephine Salmons, E. G. Izod and his son Pat, Professor Robert Young, Mr. De Bruyn, and A. E. Spiers—all played pivotal roles in bringing an attractive paperweight to the attention of Professor Raymond Dart, who in turn insightfully announced its significance to the scientific community. Science historian and philosopher Thomas Kuhn once wrote: "An apparently arbitrary element, compounded of personal and historical accident, is always a formative ingredient of the beliefs espoused by a given scientific community at a given time."[7] Such was the story of Dart and Taung. Though Dart's "discovery" was largely a matter of chance, scientific ideas about the mode and cause of human evolution rested squarely on this historical contingency for many decades. But not without controversy.

The Orphanage

Back at Buxton, things remained quiet. Not that the continued blasting of the limestone tufa did not resound across the escarpment, but little attention was paid to any further fossils that may have appeared. One might have thought that Raymond Dart, with a desire to establish the verity of his claims that the Taung child represented an ancestral human form, would have made repeated trips to Buxton to look for more evidence. But Dart was busy building a medical school in Johannesburg. It appears that he did not even visit Buxton until 1947, twenty-three years after Mr. De Bruyn had picked up the skull from quarried rubble.[8] Meanwhile the scientific community dwelled on the possible significance of Dart's child, and the peculiar dusty place from which it had been recovered.

Occasionally scholars did visit Buxton following Dart's revelation of the remarkable child's skull. Aleš Hrdlička, founder of the *American Journal of Physical Anthropology*, was the first scholar to venture to Taung, braving the strong and dusty August winds of 1925. He published a few obscure notes about his visit and the few fossils he collected, but the cave deposits from which the Taung child had been recovered had apparently already been blasted away by continued quarrying.

Research at Taung continued only sporadically. Perhaps the first serious attempt to reveal the nature of Taung was made by scholars from the University of California. Ironically, the team was first shown the site by Raymond Dart during what was putatively his first visit to the Buxton Limeworks in 1947. A geologist named Frank Peabody stayed for a year, conducting the definitive research on the limestone tufas and excavating fossil baboon bones from the spot where Aleš

Hrdlička had found similar specimens so many years before. Peabody searched vigorously for the immediate relatives of the Taung child with the courtesy of the quarry manager, who used dynamite to blast out blocks of fossiliferous rock for him. But as Peabody's field notes reveal over and over again, "No sign of Aust."[9] The Taung child, type specimen of *Australopithecus africanus*, appeared to be destined to remain a scientific orphan forever. The mystery about its place in evolution lingered, but interest waned as discoveries across Africa began to eclipse the significance of Taung.

A Time to Gather Stones Together

There is something in each and every scientist that abhors the perpetuation of mystery and myth. The unknown is not a fearsome entity to be left alone but something to be pursued with vigorous curiosity. Among the great unknowns, the question of human origins interests most people, whatever their view of life. Framed in terms of human evolution it also intrigues most scientists, and is a full-time research pursuit for many anthropologists such as myself. It was the mysteries of Taung that lured me to South Africa. What can a lone child's skull, found at the edge of the desert, tell us about ourselves, our origin, our ultimate fate? Only by building a detailed story around the skull, on the basis of fossil evidence, could the Taung child be fit into the grand picture of human evolution.

Professor Phillip Tobias, Raymond Dart's successor at the University of the Witwatersrand and a paleoanthropologist of unrivaled repute, provided me with the opportunity to satiate my own curiosity about Taung. He gave me a job in the "Wits" University's anatomy department and showed my colleagues, students, and me the precious child's skull. Then, in his own unique fashion, Tobias commanded that we go forth and multiply the number of early hominid fossil specimens from Taung.

We tried in earnest for seven years. Unfortunately, neither kith nor kin of the Taung child ever emerged from the red sandstone remains. Yet what we found was no less important for the study of human evolution. We learned what life was like for the Taung child, so very long ago, and what life is like for the people and animals who inhabit the area today. These were important clues to understanding the context of what makes human evolution tick.

Unaware of what our excavations would produce, we started off to Buxton in 1987 and began digging in earnest by 1988. As Raymond Dart wrote: "It is an area which impresses one, by its heat in summer, and its cold in winter, and by its lack of rain, of woods, of grass, of running water, and all things delectable, as the most inauspicious spot for man's forerunners."[10] This alone was incentive for investigation: why would our earliest ancestors have lived in such a place? Or *was* it such a place long ago?

Although Taung is the Setswana word for Place of the Lion, no lions were there to worry our daily and nightly activities. Lions had long since disappeared from the region as their prey became sparse in this dry and desolate land. When we arrived at Buxton, we could see that the misfortunes of climate were not the only source of devastation. It was an inauspicious spot for anything one could call humanity.

The Taung district was then part of the fictitious nation of Bophuthatswana, a "black homeland," part of white South Africa's "grand apartheid" scheme of separating blacks and whites. When Bophuthatswana was declared independent in 1976, the Buxton Limeworks closed. The timing was no mere coincidence. White South Africans would not want to live in this new "nation," and in any case the quarry was near the end of its economic viability. As the quarry managers retreated from Buxton, they left behind their workforce. By virtue of their black faces these workers, and indeed the residents of Buxton and all of Taung, were now to be citizens of Bophuthatswana. They were stuck at the edge of the desert without an economic infrastructure, without arable land, without much hope.

More than a decade later, when our team first arrived in Buxton, we were ready to find another "missing link" in human evolution. This land of climatic and political devastation waited full of promise for the discovery of scientific fortunes. And on the land waited a host of black, smiling faces, hiding their anxieties and mustering every appearance of strength in hopes of getting a job—unaware of quite what we had in mind. These villagers had survived the decade with very little, as so many of their forebears in this desolate place had done since the origin of humankind. The people of Buxton had demonstrated the resilience of human adaptability and, as I was soon to learn, exemplified the most wondrous powers of the human spirit. Together we got to work.

We had to find our needle in two haystacks—large quarry remnants avoided by the limestone company due to fills of red sandstone

Figure 3.3 • The Hrdlička Pinnacle (*foreground*) and the Dart Pinnacle (*background*), quarry remnants of the Buxton Limeworks, Taung.

(fig. 3.3). One such remnant, known as the Dart Pinnacle, was in our estimation nearer to the spot where Mr. De Bruyn had blasted out the Taung skull in 1924. About fifty meters to the east, farther away from the original escarpment, was the Hrdlička Pinnacle, where Aleš Hrdlička had recovered fossil baboons in 1925. We started work on the Hrdlička Pinnacle, as it was more accessible and much safer for digging. Moreover, at that time we believed that the sandy, fossiliferous deposits of the Dart and Hrdlička Pinnacles were all part of the same extensive cave system. We were wrong, but we did not discover the error for a few years.

Initially our team of students and locally hired workers started chipping away at the face of the Hrdlička Pinnacle, trying to remove the caked-on lime dust that had blown up from the lime kilns and masked the underlying red cave fills. Indeed most of the quarry was covered with lime dust, some of which had blown into massive dunes and drifts, giving the whole area the eerie appearance of a lunar landscape.

Meanwhile we also started looking through the dumps. Employees of the limeworks had discarded chunks of the menacing sandy cave fills they had dislodged around the quarry, forming huge piles of reddish rock. It was in these dumps where we made our first fossil discovery. We started off small—very small. In one beautiful, pink piece of rock was a rodent's tooth. Just one tiny rat incisor. At that stage it was enough to raise hopes of finding more fossils, yet in itself it was an unconvincing reward for students who had traveled to the edge of the desert looking to make great finds.

Soon a team led by my Belgian colleague, Michel Toussaint, had largely cleared the face of the Hrdlička Pinnacle of lime dust. Behind the white veneer was lovely red and pink stone, the solidified cave infill of sandstone and bone called breccia. With just a little chipping through it, they exposed a bit of tooth enamel. Careful chipping with finer instruments revealed an entire shiny tooth, much larger than the now infamous rodent incisor. Attached to that tooth was a skull. We had discovered our first fossil baboon skull, an extinct ancestor of the baboons that wander across the escarpment today.

It is difficult to describe the feeling of discovery. A rush of adrenaline accompanies pensive anticipation of what the fossil may be. Every time one of our Buxton workers came to me with a piece of breccia and said, "Mr. Jeff, it's a teeth," I felt the same excitement. The bit of enamel peeking out of the breccia could be the Taung child's long-lost brother, mother, aunt, or any remote relation. But even when, time after time, it turned out to be the tooth of yet another baboon or some other beast, the excitement persisted. Each fossil represented a link to the past, to animals that may have lived with the Taung child. These ancient and extinct creatures—baboons, antelope, and carnivores— shared resources, perhaps competed with or even ate our early ancestors, and thus helped to shape the course of the future. They were part of the initial conditions of human evolution.

Each year we went back to Buxton during the winter, when the days were not insufferably hot. The nights were clear and sometimes bitterly cold, but with the moonlight reflecting off the white lime dust, we could easily see without the aid (or hindrance) of artificial light. One could enjoy reading under the light of the full moon shining through the clear, dry air. When there was no moon, it was possible to cast a shadow even from starlight alone by waving one hand over the other. In all these different lights, day and night, we looked for fossils. Sometimes the softer illumination at night revealed a subtle visual cue we had missed in the harsh light of day: the location of some tiny fragment of tooth or bone. As the growing excavations of the Hrdlička Pinnacle took shape, they began to elicit evidence of the ancient caves and the life that had abounded millions of years ago. But the enigma of the Taung child only became more complex, for none of his immediate relatives emerged among the thousands of fossils we accumulated.

Pipe Dreams

Despite the elusiveness of the much-prized hominid fossils, some crucial answers to our questions were forthcoming. One was the time period in which the Taung child's skull was deposited. Cave deposits are notoriously difficult to date with precision. The sophisticated geochronological techniques used in East Africa, where layers of volcanic ash provide radiometric gauges of time for the fossil beds they cover, are not applicable to the haphazard accumulations of bones in the caves of southern Africa. Within the caves there are rarely any sediments or materials that could provide geological time markers.

The geological age of Taung cave deposits had long been a matter of speculation. The renowned South African paleontologist Robert Broom, despite being Dart's greatest supporter for the interpretation of the Taung child as a representative of an early human ancestor, initially suspected that the fossil was but ten thousand years old. This recent date encouraged Arthur Keith in his bid to dissuade acceptance of the Taung fossil as an early hominid, a primitive member of the taxonomic family of humans (hence his snide comment about William the Conqueror and the Sussex peasants). But without hesitation, Broom suddenly reversed his opinion on the basis of the extinct baboons presumably found in association with the child, and stated that the fossil may have belonged to the Miocene epoch (now known to have ended around 5 million years ago). Robert Broom, for all his brilliance, was not what one would call a careful scientist.

Further assessments through the years fluctuated dramatically, largely at the whim of Robert Broom. But with discoveries of *Australopithecus africanus* at other sites, most researchers accepted that the Taung fossil belonged to the late Pliocene, somewhere between about 3 and 1.5 million years ago.

The problem with the later date was that *Homo* had already evolved by then. If the species represented by the Taung skull was indeed that recent, it could not have been in our direct lineage but would belong instead to an offshoot of human evolution. If this were true, Taung would be set aside like the peasants from Sussex and left as a mere evolutionary curiosity.

Our contextual evidence placed the Taung child, or at least his surviving relatives, back in the time frame that could include *Australopithecus africanus* in our family tree. Every fossil species of animal we recovered from the Hrdlička caves was consistent with an early age. Such faunal dating, using extinct species as time markers, placed the deposits around 2.5 million years old.

Yet even with the excavation in full swing, we still did not have an adult hominid from Taung. No sign of *Aust.* To build a better case, we needed more than context; we needed the hominid itself. So we began to search the area around our excavations for other cave deposits. In particular, we looked at the quarry floor. The base of the quarriers' excavation had been a matter of pragmatics, not science, and much of the tufa and breccia below our feet remained untouched. It made good sense to investigate that base—after all, paleontologists dig down, don't they?

But it was not easy to get to the quarry floor. On top of the rocky debris left behind by the miners, a thick coating of lime dust, blown up from the kilns, had settled into a soft cement. Covering that was a layer of imported gravel forming the base of a monument that allegedly marked the spot of the Taung skull's discovery. I softly cursed the monument and its foundation, all of which came between us and the fossil deposit that might exist in the quarry floor. Knowing that heavy machinery had created this impediment to our progress, we decided to bring in a bulldozer to eliminate the problem. It was not a standard excavation technique from the textbooks but our own matter of pragmatics. Much of the work we do is decidedly low-tech. Digging is one of them.

It was great fun watching the bulldozer scrape away the gravel. The imported rock and soft cement were also cleared, leaving behind the layer of blasted debris on the quarry floor as it was in the 1920s. Indeed, paleoanthropology is one of the few professions in which "losing ground" is a *good* thing.

We then began our task, carefully sorting through every bit of rock in the remaining debris, for it had been created from the blasting of the tufa in that very area. We found breccia, we found tufa, and underneath we found the jagged, blasted surface of the quarry floor. But we did not find the fossil deposit we had hoped for. Not for two more frustrating years, anyway.

I found when I was in the field, pondering the excavation while imagining the past, that smoking a pipe was rather relaxing and enjoyable. My pipe smoking started more or less as a joke, as a play on the traditional image of the field anthropologist, but quickly I learned to savor the moments when I could smoke and let my mind wander. To alleviate the frustrations of finding nothing in the quarry floor, I lit my pipe and strolled around the excavation.

Surely we were missing something. There had to be other fossil deposits. Wandering past the confines of the carefully gridded excavation area, I went to the top of the Hrdlička Pinnacle and peered at the quarry floor below. As I precariously leaned over the edge to get a better view, in hopes of seeing more clues, and then leaned a bit more, my pipe fell out of my mouth, took one bounce off the rocky edge of the pinnacle, and landed ten meters below on the quarry floor.

I know that science normally works by reason, but when frustrated by a month of excavation at the inhospitable desert edge, one often resorts to unscientific means for inspiration.[11] The position of my pipe on the quarry floor was just that. I went down to look at the spot, an area that had not yet been totally cleared. Having nothing to lose, I decided without pause to dig there myself. Setting the pipe aside, I pulled out a few pieces of red breccia that had no apparent sign of fossil bone. But upon breaking one piece open, one little bit of breccia about the size of a golf ball, I found a tooth.

Pipe dreams indeed! It was a perfect example of Thomas Kuhn's "apparently arbitrary element, compounded of personal and historical accident." This accidental find was a deciduous tooth of a bovid, a young buck. The story it told was little different from that of the few bovids dug out of our main excavation—a young animal, grazing in the ancient grasslands, somehow lost its life and left just a touch of its remains to posterity. The small deposit from which it had come was quickly exhausted without yielding any more fossils. However, we were now assured that the idea of exposing the quarry floor might prove worthwhile, that the pinnacles were not the only sources of fossils. Serendipity and persistence, not logic, had led to this quaint discovery.

At Taung, persistence was the name of the game. Our team continued to clear the quarry floor, meter by meter, inspecting every bit of rock. Nothing escaped our attention, but all we initially found were little rodent teeth or the occasional snail shell, interspersed between sectioned sheets of tufa. Meanwhile we recovered more fossils from our excavation of the Hrdlička Pinnacle deposits. Most were fossils of extinct baboons, representing only two species. Along with those fossils we found some bovid bones and teeth, as well as the teeth and bones of a few other animals. Slowly the walls of the ancient cave were emerging, and within them were clues to prehistoric life outside the cave. Most of the animals we found were characteristic of grassland savanna environments: grazing bovids, including one not unlike the wildebeest of today. Could this be representative of the environment in which *Australopithecus* lived? Dart had certainly thought so, and now we could test his ideas.

Last-Ditch Effort

Many people who have visited my Taung excavation over the years have commented to me: "You must have the patience of a saint." Although I am no saint, patience eventually paid off. Our careful and largely fruitless exposure of the quarry floor led us west, toward the escarpment, around the cursed monument (underneath which may still be untold treasures), to the foot of the Dart Pinnacle. As we moved west we uncovered layer after layer of tufa, representing year after year of deposition. We were moving back in geologic time, across a horizontally stratified tufa accretion.[12] And at the base of the Dart Pinnacle, just before the quarry floor plummeted down to a lower level of the quarriers' excavation, remained a substantial pocket of pink and red breccia. It was another cave deposit. It was smaller than that of the Hrdlička Pinnacle, but it contained critical, earlier fossils more likely to be associated with the exact time of death of the Taung child.

These deposits were dubbed the Dart deposits (fig. 3.4). Initially various scraps of bone and fragments of tortoise shell, some rodent teeth, and the occasional baboon tooth were all that came to light. It was on one of our night tours of the site, this time with the bright moonlight belittling the amber glow of my pipe, that something unusual in the rock revealed itself. At night it was just a curiosity, a place where the smooth consistency of the breccia was interrupted by something no larger than half of my little finger. It could have been anything, but in the sunlight of the next day it was clearly a bit of bone.

The bone was nothing unusual until a bit of fine chiseling revealed that it was a cross section of a mandible. The first bit developed from the rock had a beautiful molar tooth; from the side it was not unlike

Figure 3.4 • Work on the "Dart deposits," at the base of the Dart Pinnacle, near the original discovery site of the Taung child
(denoted by the pyramid).

yours or mine, and it raised our level of eager anticipation. But every year we found something that initially had the appearance of an early human. When a fossil is encased in rock, with only bits exposed, one does mental gymnastics to reconstruct the rest. But in the field one often sees what one wants to see. Only with further development of the fossil does the true anatomy become clear, and in this case the hoped-for hominid turned into yet another baboon.

Our next discovery, seemingly humble and insignificant, was a yellowish eggshell that caused considerable jubilation. It told us very little about the prehistoric past, for it was hardly surprising that an owl or some such bird would have occupied one of these caves as they occasionally do today. Still it never ceases to amaze me that nearly 3 million years after a bird hatched out of its egg, the delicate shell could be perfectly preserved and encased in rock. The significance of this eggshell, however, was historic and not prehistoric.

When the Taung child's skull had been found in late 1924, the quarry workers spoke of finding eggshells in that particular area. In our own excavations we had not found any such eggshells and dismissed the reports as mistaken identifications of egg-shaped limestone nodules that are common in these deposits. But when we found identifiable eggshells, what sprung to mind was Hrdlička's description of the original eggshells associated with the Taung skull: they were "of yellowish uniform color and of about the size of a goose egg."[13] Hrdlička could have been describing exactly what I held in my hand. It was the first good evidence that we had finally found the very cave system from which the Taung skull had been recovered.[14]

Soon, however, our excavations exhausted the Dart deposit's cache of fossils with not even a hint of *Australopithecus africanus.* We had not found the supposed prize—the adult fossil bones. But we had found many small contextual pieces, the sum of which added up to more than their component parts. The deposit was older than those we had dug out in previous years, but the emergent picture was similar. Different, more primitive animal faces told the same story of the savanna environment in which the Taung child himself met his end.

Savanna Origins?

Back in 1925, Dart had a predilection about Taung that has strongly affected our views of human origins until quite recently. He reasoned that the ancient environment of Taung was as dry and dusty as it is today. This was based on a notion, grounded in the geological evidence of his time, that the African environment had not changed much for tens of millions of years. He took his idea further and implied that *Australopithecus africanus* must have had special adaptations to move out of the forest and trek across large expanses of treeless African plains to get to Taung. Wrote Dart in 1926, "It is obvious, *prima facie*, that the Australopithecoid group which forced this barrier into the remote Southland had evolved an intelligence (to find and subsist upon new types of food and to avoid the dangers of enemies of the open plain) as well as bodily structure (for sudden and swift bipedal movement, to elude capture) far in advance of that of the slothful, semi-arboreal, quadrupedal anthropoids."[15]

The "savanna hypothesis" of human origins was born. Dart's notion built upon a comment by Darwin, who mused that our early ancestors, like our ape relatives, "were arboreal in their habits, frequenting some warm, forest-clad land."[16] Dart's bipedal child from the dry lands of Taung seemed to represent the next chapter in our evolution. It implied that a dramatic change in morphology was related to new environmental adaptations. This view, since modified with knowledge of the climatic changes to which Africa was subjected, holds that our early tree-clinging ancestors were compelled to become bipedal, to stand up on two legs and walk, because the forests were shrinking as the African continent became drier and colder. As their populations shrank, they were forced out of the trees. Somewhere, somehow, a small group managed to adapt to the dry grassland savannas by striding into the open on two feet.

Back in Johannesburg, having washed the lime dust out of my hair and coughed most of it out of my lungs, it was time to put the pieces together from the excavation, reconstruct a picture, and build a story— to go from the rock to the hard place. Extracting bones from their rock matrix is a slow, tedious process, so much of the work will continue for some time. At present we have hundreds of identifiable fossils representing a host of ancient animals that shared the landscape with *Australopithecus.*

This assemblage of fossil fauna suggests that the Taung child and his family spent much of their lives in the savanna. Certainly there was water at Taung; without water, tufa carapaces would not have formed, and the Taung child could not have been washed to the back of a cave. Even today there is a perennial water source forming the Thamasikwa River that runs past Buxton, despite its location at the desert margin. But the types of animals we recovered from the Taung fossil deposits are those that were adapted to a savanna environment. We have grazing species of buck, not species adapted to browsing in trees. The ubiquitous baboons were small terrestrial animals as well. Even the rodents, whose tiny teeth did little to stir my students' imaginations, were clues to the past aridity. Taung may not have been as dry and dusty as it is today, for some of the fossil mammals suggest a limited presence of trees. But the ancient carnivore that dragged its food into Hrdlička's tufa caves found a lot of savanna animals nearby. Even the qualities of the tufa formation that encased both the Hrdlička and Dart caves are indicative of a dry, savanna-like environment when the Taung child's carcass drifted back to its grave. Raymond Dart had been right about the Taung environment, albeit for the wrong reasons.

But something was wrong with the picture. South Africa, indeed the entire African continent, has many environments today. Lush forests, swamps, woodlands, savannas, and deserts abound in a mosaic of ecological diversity, all inhabited and exploited by modern humans. Perhaps in the past, even the distant past, our ancestors wandered across a variety of terrains, lived in a variety of habitats. After all, the savanna was not all that inviting. There was more to eat in the woodlands and more protection from predators as well. The savanna was and still is a dangerous place for an exposed, upright creature. The mobility of *Australopithecus*, like our own slow gait, was a far cry from what Dart imagined, and certainly not capable of "sudden and swift bipedal movement, to elude capture."

Dart and those who followed may have seen only what they wanted to see in the fossils of Taung and elsewhere. Could a small population of hominids have been forced out of the forest onto their tottering two legs and feet, or were other catalysts of evolution at work? A rich fossil cave site to the north is helping us answer the question.

The Makapansgat Valley

On top of the green rolling hills, nearly four hundred miles northeast of Taung, a sentry baboon sits on a rock and casts his gaze across the trees and aloes of the landscape, ready to bark warnings to his troop should danger approach. One by one the baboons take turns at the cool waters of the stream for an early morning drink. The members of the troop, numbering one fewer than they did the previous evening, approach the water carefully before slaking their thirst. Through a series of mutters, grunts, and clicks the baboons send audible messages to keep track of one another in case of danger. A wobbly, pink-skinned baby scampers about under the watchful eye of its mother. As it begins to wander too far and the troop begins to saunter off over the hills, the mother scoops her baby up to her underbelly, where it hangs on for dear life to the mother's hair, its grip amazingly strong for an otherwise uncoordinated infant. The troop moves on from the stream in peace, though ever wary of potential threats. But as the sun comes up, only the birds and the breeze break the gentle silence of the Makapansgat valley below.

The stream fills deep, clear pools in hollows of rock that in the afternoon will refresh and entertain my students, weary of excavating through hard fossil deposits. From the pools, the water dives over the red cliffs to the relict forest below. Black eagles, whose scraggly aerie rests on a ledge midway up the northern face of the cliff, catch a thermal rising as the sun begins to warm the rocks and glide westward through the valley in search of an unwitting meal. Once their silhouettes become clear, high in the sky, a young juvenile baboon shrieks with terror, knowing he could be a target for the eagle, and runs for shelter in the nearest thorn tree.

Down below, the relentless stream of water continues to carve through the rock and nourish the forest. Vervet monkeys, smaller nimble cousins of the baboons, chatter as they feed from the branches of the trees and the valley floor, scampering from one to the other with truly acrobatic skill. In parts of Africa these vervets are known as "savanna monkeys" for the lives they lead in the vast grasslands of the continent, but here at Makapansgat they opt for the shelter and abundant resources of the forest. The forest is rich with life, from rare ferns and liana vines to a host of insects, many still unknown to the world of science. Each species hangs on to its tender spot in the intricate balance of the valley's nature.

As the valley widens toward the rich farmland of the plains between the hills, huge caves reach back into the dolomitic hillsides. These caves harbored life for millions of years. They formed through the ages as cavernous hollows deep within the earth, when groundwater dissolved large pockets of the Precambrian dolomitic limestone.

Unlike the small tufa caves of Taung, massive caverns formed here. With the uplift of the African continent, the surface erosion that cut the Makapansgat valley eventually opened the caves to the air, providing a deceptively inviting refuge for the animals that dared to venture in. Along with a host of diverse creatures who found either life or death within the Makapansgat valley caves were humans and their ancestors.

One such cave, or *gat* in Afrikaans, was the frightful scene of the 1854 siege of the Ndebele chief Mokopane and his people. Nearly two thousand Ndebele had taken refuge in the dry, expansive cave, trapped by an Afrikaner brigade of militia who sought territorial expansion. All decided to fight and perhaps die for the riches of the land. In the end, after nearly two weeks of anguish, it was the Ndebele who died of starvation and dehydration in the cave, or were massacred upon surrendering.

One can still find human remains littering the cave, testifying to the tragic deaths of well over a thousand Ndebele men, women, and children. And underneath the few remaining bones and artifacts of the Ndebele people of 1854 lies nearly half a million years of cave sediment recording their ancestor's successes and failures at adapting to life in the Makapansgat valley. Indeed, the valley is as rich with prehistory as it is with the diversity of life there today.

The valley forest below the infamous cave provides a luxurious habitat for a more peaceful, distant relative of the humans, baboons, and vervet monkeys: a wide-eyed nocturnal creature known as a bush baby, or galago. During the dark hours of the previous night, despite considerable weariness, student scientists and their mentor roamed the starlit valley with bright artificial lights to flash into the eyes of the equally curious galagos, who sprang from tree to tree in search of insects for sustenance. The galagos were probably more aware of the danger that lurked in the dark valley than we humans.

That particular night, as we returned to camp, a thunderous ruckus erupted from the escarpment on the north side of the valley. From the sounds it was clear that a leopard had attacked a young baboon and aroused the protective behavior of the troop. For twenty minutes or so this intense battle of nature ensued. The shrill cries of the cat and the boisterous barks of the baboons echoed through the valley while the most "advanced" primates in the valley pondered our all too recent defenseless walk through the wilds, easily within attacking range of the hungry stalking leopard. It was a moment to savor, to contemplate the fragile human place in the chain of being, as we huddled around the dying campfire, the discarded bones of our own meals simmering and blackening in the coals. We threw another log onto the fire for warmth and a semblance of protection.

In the light of day, my tape deck booms Vivaldi's *Four Seasons* against the hills. Amid classical echoes, students search for physical remains of the events of the previous evening. A few broken branches and scattered leaves are all they find. The baboon that fell victim to the leopard was perhaps stashed in one of the nearby caves or hung in a tree somewhere, out of reach of the jackals and other scavengers of the valley. We rekindle the fire for breakfast over the ashen white bones of the Saturday night feast. It is another Sunday morning in the Makapansgat valley.

Promethean Origins?

Over the top and down the other side of the hill that houses the infamous Historic Cave of Mokopane, or Makapan as he became known, are the remnants of considerably older caves. Deep within, the caves accumulated clues to a variety of ancient stories. The older caves of Makapan's valley have long since filled with limestone stalactites, stalagmites, and the debris of ancient times, including fragmented bones of prehistoric prey animals, dragged in by predatory beasts. With time, the cavernous openings filled to a point where nothing else could enter, and the contents inside were naturally sealed—that is, until humans, once again in search of limestone, blasted their way through the ancient deposits, now solidified into rock (fig. 3.5). There were plenty of giant limestone formations within the caves for the quarriers along with, once again, tons of economically worthless sandy rock, filled with scientifically priceless fossilized bone.

In 1925, when Raymond Dart startled the world with his revelation of the "ape-man" child from Taung, he captured the imagination and interest of many people across the globe and a few would-be excavators. One such person was Wilfred Eitzman, a mathematics teacher in the town of Pietersburg, just north of the Makapansgat valley. Eitzman visited the quarry operations at the Makapansgat Limeworks where, as was the case at Taung, the sandy cave breccias, poor in lime, were tossed aside into massive waste dumps. Having had his curiosity

Figure 3.5 • Quarried cave entrance at Makapansgat Limeworks.

piqued by Dart's discovery at Taung, Eitzman perused the blocks of breccia and readily found fossilized bone.

Eitzman's fossil bones from the Makapansgat Limeworks were remarkable for their dark, black color. He sent them to Dart, who promptly published his observations as "A note on Makapansgat: A site of early human occupation."[17] Although there were no early human fossils or stone tools from the site giving evidence of human occupation, Dart deduced that the blackening of the bones was the result of fire. Like the bones we discarded at the edge of our fire hearth back at the camp in the valley, these ancient bones had apparently become carbonized. It was clear to Dart that some early hominid occupants of the cave had used fire. Indeed, initial tests of the fossil bone revealed the presence of free carbon, a sure sign that the bones had been burnt.

As was the case at Taung, the many other fossil bones at the Makapansgat Limeworks remained unnoticed and unstudied until many years later. Scientific excavation finally began in 1948 under the direction of Raymond Dart , whose interest in paleoanthropology had been rekindled. Dart relied on the remarkable "X-ray eyes" of famed fossil finder James Kitching, a man, like Robert Broom, who had a penchant for finding fossils of the dinosaurs and mammal-like reptiles of the South African karoo. Kitching did not take long to find the back of a skull of a peculiar creature, embedded in a gray breccia. The cranium represented by this occipital bone was undoubtedly that of *Australopithecus*.

By the late 1940s there were a number of dolomitic caves in the Transvaal that had yielded *Australopithecus*, mostly discovered by Robert Broom, Dart's bulldog. The fossils confirmed Dart's notion of the creature being an upright, small-brained human ancestor. But none previously had been associated with fire. Dart thus named a new species from Makapansgat, *Australopithecus prometheus*, after the Titan from classical Greek mythology who brought fire down from the gods to mortals on earth. The modest brain of *Australopithecus*, slightly larger than that of a chimpanzee, appeared to have been sufficient to capture the power of fire.

The blackened fossils of the ancient cave included remains of over forty large mammal species as well as an assortment of smaller rodents and bats. Dart suspected that all these bones had been left behind by *Australopithecus prometheus*, who must have been a formidable hunter more than 3 million years ago. Indeed, in Dart's eyes, the broken and blackened bones of the hominids themselves, appearing sparsely throughout the breccia deposits, could only have been left behind by the beast that controlled the fire. A picture began to emerge of our ancestors as vicious, probably cannibalistic killer ape-men. The idea was certainly consistent with the evidence over the hill in the Historic Cave, where modern humans wrought carnage and devastation upon each other, but only the fossils from the limeworks could tell us if such evil ways had a very early start.

Daggers I See before Me

Dart's careful inspection of the thousands of fossil bones recovered from the Makapansgat Limeworks cave revealed intriguing disparities in the massive assemblage. Certain bones were much more common than others. The mandible, or lower jaw, was the most frequently found body part. Perhaps, Dart thought, the mandibles, fitting neatly into the adroit hands of the bipedal hominid, would have made effective saws or scrapers.

The distal halves of the humeri, the elbow end from arm bones of the hominids' supposed victims, were also found in great abundance. Usually the shaft had been broken in a spiral manner, leaving an effective and sharp blade tool. Could this be the tool used by *Australopithecus* to kill its prey? Indeed, many bone fragments were found that would have made effective daggers. As Dart cast his perceptive eyes across the fossil bones, he set his imagination free. The bovid horn cores—the bone that filled the keratinous horns of the extinct antelope—would have been effective for defense when wielded in the hands of *Australopithecus*. There were all sorts of potential bludgeoning tools and sharp knives, he decided, comprising the weaponry certainly used by the predatory ape-man.

As if the kitchen cupboard of our Promethean ancestor had been opened by the lime miners who gutted the contents of the ancient Makapansgat cave, Dart recognized a plethora of domestic implements. *Australopithecus prometheus*, according to Raymond Dart, was a predatory beast dependent on tool culture. Dart christened this early ancestral achievement the Osteodontokeratic Culture. *Osteo* means bone, *donto* refers to teeth, and *keratic* alludes to horns. The ODK, as we now shorten the term for obvious reasons, was an alleged bone-, tooth-, and horn-tool culture that predated the earliest stone-tool culture by more than half a million years.

The apparent predatory nature of Dart's Promethean ancestors sparked interesting speculation about the aggressive nature of humankind. The imaginative Raymond Dart himself penned lurid descriptions to elucidate the behavioral implications of the ODK Culture: "On this thesis man's predecessors differed from living apes in being confirmed killers: carnivorous creatures, that seized living quarries by violence, battered them to death, tore apart their broken bodies, dismembered them limb from limb, slaking their ravenous thirst with the hot blood of victims and greedily devouring livid writhing flesh."[18]

Let it never be said that Raymond Dart did not have a way with words. Reread Dart's words aloud, with feeling, and you will understand what I mean. When asked why he used such vivid language, Dart simply answered: "That will get them talking!"

Despite the riches of bones at Makapansgat, fossilization is a rare phenomenon, as I certainly discovered at Taung. A lot of peculiar things have to happen in just the right way before an animal can contribute part of its skeleton to a grave that will be preserved in the fossil record. The science of taphonomy, quite literally the study of the grave (*taphos*), is a necessary preoccupation for paleontologists trying to interpret South African caves. We cannot know what went on in the surrounding environment until we recognize the forces that brought carcasses and their bones into the cave.

Thomas Huxley is oft quoted for writing that the "tragedy of science" is the "slaying of a beautiful hypothesis by an ugly fact."[19] The weight of the facts of nature, of the activity of carnivores and porcupines, proved that Dart had overinterpreted the fossil evidence, seeing only what he wanted to see. Charles K. Brain, better known as Bob, pioneered South African cave taphonomy. He clearly demonstrated that the supposed artifacts of the Osteodontokeratic Culture were nature-facts—the work of cats, hyenas, and porcupines. Little or no evidence remained at Makapansgat of early human tool culture, and visions of our horrific cannibalistic past faded. As Brain noted in such elegant and simple language, our ancestors were the hunted rather than the hunters.

Bob Brain, incidentally, later found true bone tools at the more recent Transvaal cave site of Swartkrans, there associated with the genus *Homo*. When Raymond Dart asked Brain what the tools were used for, Brain answered with his well-tested evidence that the bones had been used for digging up roots. "Brain, that is the most unromantic explanation I have ever heard," said Dart. Brain matched wits with Dart and continued his explanation. The aging but still vivacious Dart then got a wicked gleam in his eyes, picked up the sharpened bone digging tool and, edging it forward toward the younger naturalist, said, "Brain, I could run you through with this!"[20]

But what of the evidence of fire at Makapansgat? Certainly the controlled use of fire, or perhaps even the ability to create fire, would have given an advantage to our early hominid heroes in the Makapansgat valley. Few carnivores then or now, no matter how fierce, would dare to attack a creature possessing the flame. Even the unnatural glow from battery-powered torches probably protected my students and me on that fateful night when we were surely within pouncing distance of a particularly ravenous leopard. But *Australopithecus* probably did not have fire.

When bone is exposed to fire at low temperatures, it may turn black. The bone surface becomes carbonized. But when heated by fiery logs, like the bones in our campfire hearth, the black soon gives way to white. The black carbon becomes oxidized; bones lose it and turn chalky white. What, then, caused the blackening of the Makapansgat fossil bones? When scrutinized under a microscope, the bones reveal the answer: a mineral called manganese. The dolomite in which the caves formed is rich in manganese. On some of the bones, one can see black dendrites under the microscope, little crystal "trees" formed by the manganese. Manganese, not fire, blackened the Makapansgat fossil bones.

But Dart had tested for, and found, the presence of free carbon on the fossil bones. That was unequivocal evidence of fire. True, but the fire was not necessarily as old as the bone. Repeated tests for free carbon did not confirm its presence on all the bones. One must realize, however, that the fossiliferous breccia was blasted out of the ancient cave deposits by miners' dynamite. Recent explosions, contemporary pyrotechnics, were responsible for the presence of free carbon on the ancient fossils. Serendipity in science can lead not only to discoveries but also to false conclusions. Without scientific persistence, without the skeptical attitude that Huxley urged on us, we may see only what we want to see.

Stripped of their Promethean heritage, Dart's fossils are now considered to be the earliest southern African representatives of *Australopithecus africanus*, like those of Taung and Sterkfontein. About 3 million years ago, these early Makapansgat hominids had no sophisticated tools. Their brains, only slightly expanded compared with those of their nearest primate counterparts, may have toyed with novel ideas for manipulating the environment and increasing their chances of survival. But there is no doubt that they were the unwitting targets of an unfathomable array of beasts who saw them not as *Australopithecus africanus* but as food.

Seeing the Trees for the Forest

The Makapansgat Limeworks assemblage of mammalian skeletal pieces reveals the remains of a frightening number of carnivores. Two distinct species of saber-toothed cats prowled the area back then. We know little of their nature or behavior, but one may safely assume that carnivores of such substantial size, possessing those horrendous teeth, were greatly feared by *Australopithecus*. The saber-toothed cats existed in addition to the more familiar leopards as well as cheetahs. Lions, which make their first appearance in the local fossil record shortly after, may have been around as well.

Our ancestors in the Makapansgat valley also had to deal with two species of hyenas. These were not small, complaisant hyenas. In fact, *Australopithecus* may have been quite a delicacy, when available, for the hyena packs that crunched so many of the bones in the Makapansgat caves. And, of course, the jackals and foxes were probably not welcome sights during a hominid's daily millings about. Without tools of stone or bone, without fire, without refuge in the carnivore-infested caves nearby, it is a wonder that the slow, bipedal ape-humans survived at all.

Fortunately for the hominids, there were plenty of other animals around to satisfy the wide array of hungry carnivorous beasts and dissuade them from frequent attacks on our ancestors. Baboons, for example, abounded. Most came in one of three sizes—small, medium, and large—perhaps representing three different species but certainly signifying high baboon diversity. A fourth type, certainly a separate species that specialized in eating grass seeds, occasionally wandered through as well. The baboons were accompanied by a large leaf-eating monkey that lived above them in the trees. Of all these primates, *Australopithecus* is the least frequently found in the Makapansgat fossil deposits, indicating that the ancient carnivores may have preferred to dine on the hominid's monkey cousins, or at least that they found them available on the menu more often.

In my own excavations of the Makapansgat breccias I recovered numerous baboon skulls. On the basis of my finds one could conclude that the carnivores, the taphonomic agents of bone deposition, focused their attentions on these small-brained primates. But largely by chance, my work has concentrated on a particular part of the cave that is baboon-rich. As at Taung, finding fossil baboons seems to be my paleontological destiny. Other researchers before me, tapping other parts of the early cave deposits, have found the rest of the prehistoric story. The Makapansgat deposits, it appears, became the final resting place for an astonishingly diverse assemblage of animals. This fossil site is now one of the richest known for species diversity, preserving more large mammal types from the past than exist in the area today. Not just primates and carnivores but assorted species of many families indicate that a high degree of biodiversity existed in the valley more than 3 million years ago.

Imagine the contented life of a saber-toothed cat on the prowl in the primeval Makapansgat valley. Baboons to satisfy appetites large or small, the short-necked giraffe, and two or three pig species could have started the menu. If a cat had a taste for venison, there were innumerable antelope browsing on the leaves of the trees, with more choices at the edge of the forests. The cold scientific name *Makapania broomi* hardly reflects the warm flesh of one such prehistoric beast; but to scientists in the know, the name elicits grand images of a substantial, musk ox–like bovid that munched a variety of the rich Makapan valley vegetation and, once plumped up, unwillingly served itself for dinner.

Of course other carnivores such as leopards joined the sabertooths. They must have left the Makapansgat valley heavily littered with carcasses for the scavenging beasts, the jackals and hyenas, which could have lived quite comfortably if their own pack hunting had been insufficient. Much of this booty could have been taken back to the nearby cave to feed a litter of young, where year by year the bone-crunching teeth of the hyenas created an auspicious collection of curiously damaged bones for posterity. Little of what remained in the valley was wasted. Thousands of species of plants and animals worked together, it seems, serving nutrients to all parts of the food chain to produce a measure of ecological order. A chaotic attractor, indeed.

Even the porcupine, just trying to wear down its incisor teeth and gain a bit of a mineral supplement in the process, could have found bones of anything from a hippopotamus to a mouse. The porcupine would scurry back to the far reaches of the cave with such bones. If detected en route by a passing hyena, it would raise and rattle its quills, grunt loudly, and stomp its feet until the pest went away like a frightened puppy. But the occasional leopard, undeterred by such a display, would casually roll the porcupine over and snack on its soft underbelly. The law of the jungle, so to speak, ruled the day.

Among the animals and amid the trees wandered our own ancestors, who had a somewhat different perspective on life in the forest. The nervous night my students and I endured, sharing the valley with a leopard, was only one small sample of what our ancestors encountered daily. Occasionally they would submit to the ultimate terror of being attacked, mangled, and consumed by carnivorous predators and would eventually be contributed to the bone assemblage of the cave deposits. For some unknown reason it does not *appear* to have happened often, for fewer than thirty of the tens of thousands of bones recovered from the Makapansgat breccias represent *Australopithecus*. Perhaps the early hominid was only a rare visitor to the valley, or maybe its lean body was not worth the trouble of attack by otherwise contented predators.

Fossil biodiversity implies living biodiversity in the ancient Makapansgat valley, and such rich variation in turn reveals a past environment vastly different from that of Taung. It must have been a lush, wet habitat to support the browsing bovids as well as the variety of primates and other animals that, in turn, fed the array of carnivores. Among all these beasts was *Australopithecus*, our supposedly savanna-dwelling ancestor. Although the hominids were infrequent guests in the cave, their appearances were common enough to leave behind a number of remains and suggest to us today that our early ancestors made extensive use of the forested valley.

The revelation of early hominids as occasional if not frequent forest dwellers poses a problem for the long-cherished notion of the origin of bipedalism. It was supposedly the decline of the forest, and the expansion of the savanna, that forced our ancestors onto their feet.

But it should not surprise us unduly that such a scenario is a bit too simple. After all, the vervet monkeys that love the relict forest of the Makapan valley today can be equally at home in the savannas of East Africa, hence their alternative name of "savanna monkey." Certainly our own ancestors could have been equally adept at making a home of both savanna and forest.

What is intriguing from the contrasting fossil records of Taung and Makapansgat is that *Australopithecus* survived, and eventually thrived, in a variety of environments. From an early stage of evolution, our ancestors were not just adapted in a Darwinian sense but adaptable. As generalists rather than specialists, they were poised to dominate the future landscapes of Africa and beyond.

To give Dart his due, it should be noted that he recognized the significance of the differences between Taung and Makapansgat. In an oft neglected passage, written in 1964, Dart wrote with characteristic hyperbole: "The distribution of South African sites from Taung to Makapansgat and the climatic variations to which these sites were subjected geologically during their occupation by australopithecines show that the South African types had adapted themselves to almost, if not quite, as wide a range of climatic, soil and vegetation variation as modern man faces outside the arctic circle."[21] Whereas Dart had clearly forgotten his youthful experiences in the variable climate of Ohio, the point was well taken that *Australopithecus* was not solely adapted to a specific environment. But in the minds of most paleoanthropologists the savanna hypothesis persisted—until investigations across Africa rendered the notion dead.

The Savanna Hypothesis Takes a Hike

Fossils from East Africa now reveal a very different story from that of the savanna hypothesis. What is possibly the earliest evidence comes from Aramis, in the Middle Awash valley of Ethiopia, where the Afar-speaking peoples, like the Tswanas of Taung, eke out a living in harsh desertlike conditions. There Tim White, a paleoanthropologist from the University of California–Berkeley, and his colleagues discovered a new and different species, *Ardipithecus ramidus*. One of the key specimens belying the primitive yet hominid-like characteristics of the species was that of a child, a small irony coming nearly seventy years after the discovery of the Taung skull.

The Afar word *ramid* translates as "root," an appropriate appellation for such a primitive, chimpanzee-like hominid. The seventeen fossils may indeed prove to be right at the root of our family tree, for their barely humanlike features were accompanied by an age determination of around 4.4 million years ago. They are perhaps more than a million years older than even the earliest hominid fossils known from South Africa.

Any find of early hominid remains is bound to cause as much controversy as excitement. This will no doubt be the case for White's *Ardipithecus ramidus*, just as it was for Dart's *Australopithecus africanus*. There is still much to be learned about both species in the tender young science of paleoanthropology. But the initial evidence from Aramis is worth considering here for a particular reason: it suggests a woodland environment in which these earliest possible ancestors of ours lived and died. Aramis is dry and desolate today, but 4.4 million years ago it may have been a veritable Garden of Eden.

The apparent hominid successor of *Ardipithecus ramidus* was a primitive-looking biped from East Africa going by the name of *Australopithecus anamensis*. An early relative of the Taung child, dating to as early as 4.2 million years ago, this creature had only a few of the features one could call human. But the bones show that one humanlike feature was bipedalism, making *A. anamensis* the earliest confirmed creature to walk on two legs. Did it live in the savanna, as it should have by Dart's hypothesis? Certainly the bones were deposited in sediments along a rich riverine forest, leading some to believe that the hominids exploited this rich environment. Drier and more open conditions prevailed a short distance from the river, but we do not know how often *A. anamensis* may have ventured out of the forest. Its successors, however, seem to have lived in both types of habitats.

Closely following on the heels of *Australopithecus anamensis* was *Australopithecus afarensis*. The most famous representative of this species is a partial skeleton nicknamed "Lucy," a most remarkable fossil indeed. The skeleton clearly shows the upright, bipedal nature of these early hominids. Lucy and her friends seem to have lived sometime between 3.6 and 2.8 million years ago, thus overlapping the time their closely allied cousins lived in South Africa. But as now appears to have been the case for the adaptable *Australopithecus africanus*, it has been demonstrated that *A. afarensis* also lived in a variety of environments. In Ethiopia, at the Hadar site that revealed Lucy and other beautiful specimens of these early bipeds, the diverse mammalian remains bespeak a "warm, forest-clad land"—Darwin's image of the ancient landscape that Lucy and her cousins, by now walking fully upright, should have left behind if the savanna hypothesis were correct. Ironically, one of the more lush of the *A. afarensis* habitats is in evidence at the Ethiopian site known as Maka, dating to just a few hundred thousand years before Makapansgat. Yet in Tanzania, at the *A. afarensis* site of Laetoli, drier and more open features of the landscape prevailed.

It is becoming increasingly apparent that the initial hallmark of human evolution, walking upright with a bipedal gait, predated the supposed environmental cause. Global temperature declines and the progressive decline of African forests began about 3 million years ago, long after *Australopithecus* had established its peculiar gait. Moreover, given the range of habitats apparently utilized by *Australopithecus*, no forest reduction or savanna expansion seems necessary as a causal agent. A whole new picture of our bipedal origins now emerges.

DART'S CHILD from Taung first gave us a vision of our ancestors as bipedal creatures forced into adaptations by new environmental conditions. But perhaps Dart and those who followed were just seeing what they wanted to see, with reasonable hypotheses formed out of preconceived ideas. The historical contingencies of discovery undoubtedly shaped their scientific opinions. But new light on old fossils brings novel perspectives. From Taung to Makapansgat, Laetoli to Maka, paleoanthropologists are unearthing more fossils, filling in new pieces of the human evolutionary puzzle.

The hominid status of the bipedal *Australopithecus africanus*, which so many doubted in Dart's time, turned out to be correct. The savanna hypothesis, and Promethean Osteodontokeratic Culture, now appear to be wrong. Taking shape instead is a vision of *Australopithecus* as an adaptable creature with no apparent environmental preference and no specific savanna adaptation, yet no material culture. Gone is the notion that reduced forests suddenly forced our quadrupedal ancestors to stand up and walk proudly across the savanna, for it appears that bipedalism preceded any great degree of forest reduction.

This presents us with a problem. Environmental change has often been seen as a driving force of evolution. There is no doubt that the situations in which early hominids found themselves did change dramatically over millions of years: the African continent did gradually dry up. Novel resources became available as the hominids ventured into new territories and environments. As old foes such as the saber-toothed cats went extinct, new competitors and predators evolved to challenge the very existence of the evolving hominids. A multitude of possible evolutionary catalysts and deterrents shaped the initial conditions for the descendants of *Australopithecus*—the genus *Homo*. Scientists thus must take a fresh look at the correlations, coincidences of change, to hypothesize and test the mechanisms of our evolution.

4

Speeding Up the Pace of Evolution

EVOLUTION is a slow process, at least to human perceptions. It took as much as 5 million years or more for human beings and our nearest living relatives, the chimpanzees, to evolve from our common ancestor. But in many other respects we have not changed much through all that evolutionary time. One has only to study a chimp closely to see a fuzzy reflection of oneself in both looks and behavior. Genetically we are 98 percent to 99 percent identical to chimps. Five million years seems like a long time for those few differences to accrue.

It took around 3 or 4 million years for us to evolve from early *Australopithecus.* During that extensive period of time our hominid brains became larger with our heads held more aloft, our teeth and faces shrank, and our gait became more securely bipedal. There were certainly innumerable other changes as well, of minor or major consequence. Despite the slow accumulation of unique human traits, the expansion of our brains seems to have happened quite swiftly, more than *tripling* in size throughout that time. The pace of evolution is thus a matter of perspective.

Three million years is a very long time by our own standards, but not incomprehensible. Take a look at your watch and imagine that every second represents one year of human evolution. First, count the ticks representing your own life—not a long wait, even for an impatient person. Perhaps you would like to take some more time, well over two minutes, to ponder evolution while counting back to the time that Darwin and Wallace announced the theory of evolution through natural selection.

As the seconds continue to roll by, however, your patience may wear thin. While you go back to the time of Galileo's astronomical and physical observations, over six minutes of this little game will elapse. If you want to try this exercise back to, say, the year A.D. 1, be prepared for a contemplative period of more than half an hour. The origin of fully modern members of our own species, *Homo sapiens sapiens*, was nearly twenty-eight hours ago. By this means of counting, we *were* born yesterday (merely 100,000 years ago by conventional time).

Don't even try to wait for your watch to take your imagination back to the origin of our genus *Homo* and the first appearance of stone-tool culture, some 2.5 million years ago. You would have to wait for nearly a month. Double that for the split between human and chimp ancestors.

Human evolution may be slow in our eyes, but from a geological perspective it all happened quite quickly. To wait one contemplative second for each of the earth's 4.5 billion years of existence would take approximately 142.6 years, longer than a human lifetime. Within that long span of our planet's existence, the single month of human evolution doesn't seem quite so extensive. But our millions of years still represent a long time, and evolution is still slow. Or is it?

The pace of evolution has long been a matter of considerable debate. Although we all agree that evolution is slow by the standards of a human lifetime, it is not quite clear how slow is "slow." Likewise it is legitimate to ask whether or not evolution is occasionally less slow and other times very slow. Part of our conceptual problem is one of scientific perspective: a geologist studying fossils over a vast period of time across two continents is bound to define slow differently than a population geneticist studying changes in the genes of fruit flies in Hawaii. From the perspective of a paleoanthropologist, some key questions arise: Did the pace of human evolution vary between degrees of slowness? If so, why? If not, why not?

Let me float you across the surface of the debate on the pace of evolution before I plunge you into the depths of how this relates to human origins. Darwin gave us a vision of evolution as a slow process, from the perspective of one who could not view dramatic change in species within a human lifetime. Fossils of extinct organisms in ancient geological strata showed that evolution transformed species over extensive periods of time, although gaps in the fossil record did not allow a full viewing of the slow accrual of change.

On the other hand, perhaps at least some gaps in the fossil record and the seemingly sudden appearance of new species represent real events. This perspective is the cornerstone of a recent rallying point among evolutionists known as the theory of punctuated equilibrium. The notion has been championed by paleontologists Niles Eldredge and Stephen Jay Gould.[1] Like Huxley, they can be good Darwinists and still challenge a particular emphasis of Darwin's vision.

In short, Eldredge and Gould see evolution as an uneven process. Evolving species, such as *Australopithecus africanus*, appear to be rather static through long periods of time. This lack of change, as often perceived in the fossil record, is an "equilibrium." Occasionally something happens to upset the static equilibrium, and an event of sudden change "punctuates" the development of the otherwise complacent species. Thus Eldredge, Gould, and their countless disciples do not view evolution as a gradual process of incremental change but as long periods of equilibrium interrupted by fast bouts of change. In the words of some unkind detractors of punctuated equilibrium, it is the "theory of evolution by jerks."

The novel aspect of punctuated equilibrium is not in the variability of evolutionary rates. Even Darwin noted that evolution through natural selection sometimes proceeds at faster (less slow) rates than at other times, although it was not his overriding concern. What is original and controversial about the theory is that it claims *dominance* for a punctuated mode of speciation. In other words, long stasis followed by sudden change accounts for more species origins than the gradual adaptive change often attributed to Darwinian evolution.

It may not be important whether punctuated equilibrium or gradualism more often accounts for the origin of new species. Both tempos and modes of evolution (to borrow a phrase from George Gaylord Simpson) do happen. What we need to find are evolutionary mechanisms that can account for both gradual and punctuated modes of change, as well as find when and why each occurred during our evolutionary history. As an analogy, both Beethoven and Mozart were important composers who still influence music; I do not care which has been more popular and influential, as that is a subjective matter of histori-cal judgment.[2] I want to know *why* the music of each composer was so effective, and what implications their musical offerings have for us today. Likewise, *why* does evolution sometimes proceed at the pace of a snail and sometimes leap into new realms with the swiftness of a hungry cheetah?

There is little doubt that punctuated events occur in evolution. Sometimes when it rains it pours. In an effort to understand the evo-lutionary cloudbursts, most scientists who embrace punctuated equi-librium look for outside causes to explain both the punctuation and the equilibrium. Environmental change, whether a cooling climate or the advent of a new and effective predator, is usually seen to be the driving force behind relatively quick events of evolution.

Alternatively, mammalian evolutionary novelty may be caused by intrinsic principles of each species and be only mildly *governed* by extrinsic factors. Environmental change then would not be a catalyst of species origins. Throughout this book I will explain what I mean by that. At stake is an understanding of the true causes of human evolu-tion. The origin of our genus, *Homo*, begins to illustrate the point.

Handy Man and Friends

Sometime around 2.5 million years ago the hominid lineage began to diversify. Our family tree began to branch further from its australo-pithecine bough. We see evidence of the origin of our genus *Homo*, including our ancestral species and perhaps some others. Along with the limb that eventually extended to *Homo sapiens* was at least one other offshoot leading in another direction—the "robust" australo-pithecines. Meanwhile *Australopithecus africanus* was still around in South Africa, and East Africa had a species known as *A. garhi* with some *A. afarensis*–like traits.[3] But although biodiversity increased among hominid species, only one lineage would survive to this day. The others disappeared as mysteriously as they arose.

Australopithecus robustus, a South African species with a wide, heavily built face, was an evolutionary cul-de-sac. Evidence of these creatures appears earliest in Transvaal cave deposits, such as those at Swartkrans and Kromdraai, dating to around 2 million years ago. The robust facial bones, housing enlarged teeth and supporting substantial attachments for chewing muscles, reveal a hominid focused on powerful chewing—the grinding mastication of high-fiber food. It has been suggested that the facial specializations imply a vegetarian diet for *A. robustus*. New high-tech clues from isotopic analyses of the teeth have challenged this notion to a small degree,[4] for there is possible evidence of some meat in the diet of this peculiar creature, but such supplements of soft flesh probably did not keep the hominid from a substantial helping of rough, hard food for which the chewing apparatus of its face was well adapted.

Like all hominids, *Australopithecus robustus* walked around the southern African veld with a bipedal gait. Likewise, in East Africa, the hyper-robust *Australopithecus boisei* followed in the footsteps of a proud australopithecine heritage. These cousins are related in their specialized facial adaptations, but their specific ancestry is not known. Some paleoanthropologists see them as a separate genus, so one may read of *Paranthropus robustus* and *Paranthropus boisei*. Indeed the earliest of the East African form is occasionally referred to as *Paranthropus aethiopicus*. It is probably unimportant to quibble about names. However, it is noteworthy that the bipedal beings with large faces all went extinct after having lived side by side with the earliest members of the genus *Homo* for perhaps as long as a million years.

The extinction of a species, like the death of an individual, is a normal process. Unlike death, however, extinction is not always terminal. For example, *Australopithecus africanus* is extinct, but it may well have evolved into other forms such as early *Homo* and/or *Australopithecus robustus*. Some early australopithecine traits were lost and new features gained through incessant evolutionary change, but the chain of being persisted. There was no such luck for the robust creatures who met with an unceremonious and permanent end.

On the other hand early *Homo* appears to have been quite successful. It makes its debut in the fossil record at about 2.4 million years ago but probably arose a bit earlier than that. Stone tools appeared at roughly the same time. Perhaps the expanded brain of *Homo*, so profoundly evident in the early fossil crania of the genus, gave our own ancestors a greater degree of adaptability in a challenging environment. Such a brain, with disproportionately increased gray matter of the neocortex on its periphery, may have allowed foresight, planning, and even communication.

Certainly primitive stone tools, simple but sharp, added to the behavioral repertoire of these more sapient creatures. Unlike the putative skeletal tools of the so-called *Australopithecus prometheus*, stone tools are not just the product of vivid archaeological imaginations. The artifacts are abundant, comprising flakes of stone crudely hit off one pebble by another—a toolmaking tradition known as the "Oldowan" industry. The tools did the trick, for cut marks from Oldowan tools are found on the bones of ancient animals. Members of the early *Homo* genus were not yet avid hunters eating "livid, writhing flesh" but were probably opportunistic scavengers; marks from their sharp stone tools sometimes cut across the broader, distinctive tooth marks left by the more effective carnivores who had their share first.

The more we find of early *Homo*, both in southern and eastern Africa as well as in the "hominid corridor" of Malawi in between, the more interesting the story of this lineage gets. Certainly the earliest members of our own human genus were quite variable. Relative to the australopithecines, they were all characterized by an expanded brain and a somewhat reduced and modified face. They were beginning to throw off the last vestiges of animals who habitually climbed trees, but even the fossil limb bones contain a curious mosaic of features useful for both climbing and walking. Bit by bit, our ancestors were becoming more upright and more human.

So far I have referred to early *Homo* without committing myself to any particular species name or names. In 1964, when I was a just a young child, a new species was christened, representing a link between *Australopithecus* and *Homo erectus*, the latter being the penultimate species before our own. The species was called *Homo habilis*, or "handy man." Soon thereafter *Homo habilis*, as named by the famed Louis Leakey and his anatomist friends Phillip Tobias and John Napier, became the focus of considerable controversy. Such is the modus operandi of paleoanthropology. Some saw *Homo habilis* as a late version of *Australopithecus africanus*. Others viewed it as early *Homo erectus*. That dichotomy alone confirmed that the fossils, found by Leakey at Olduvai Gorge in Tanzania, were truly transitional.

The issue had not been resolved by the time I started studying anthropology at university level in 1976. My professors led me to doubt the validity of the species. Little was I to know that ten years later I would take a job with Phillip Tobias, one of the original proponents of *Homo habilis* and the person responsible for the detailed description of the fossils. By the time Tobias finished his magnificent two-volume monograph on the Olduvai fossils in 1991, the picture had become considerably more complex. In contrast to my own efforts at Taung and Makapansgat, my fossil-hunting colleagues were finding early hominid remains all over the continent.

Because fossils of *Homo habilis* were first found in East Africa, it appeared that *Homo* arose in that part of the continent. But in 1976, at the South African Transvaal cave site of Sterkfontein, Alun Hughes discovered a representative attributed to *Homo habilis*.[5] By the time I met Hughes he was a crusty old digger, the greatest master of eliciting fossil bones from the intractable cave matrix. He had worked for Dart at Makapansgat and later worked at Sterkfontein under the direction of Phillip Tobias. Sterkfontein, of course, is a cave site known for an incredible assemblage of *Australopithecus africanus* fossils. There Hughes found literally hundreds of fossil hominid pieces, making him the world's most prolific discoverer of early hominid fossils. As at Makapansgat and Taung, ancient predators had captured these hapless creatures and dragged them along with other prey into the shelter of a cave near what is now Johannesburg.

Adjacent to the *Australopithecus* deposit of Sterkfontein is a cone of debris that filled in much later, around 2 million years ago. There Hughes carefully removed pieces of skull bone and teeth that could only be attributed to early *Homo*. In the same deposit are stone artifacts, confirming the presence of handy man. So across Africa we see evidence for the emergence of our own genus. For the first time, a single species, *Homo habilis*, was recognized from both eastern and southern Africa.

Meanwhile Richard Leakey, son of the original discoverer of *Homo habilis*, capitalized on his familial skills and found more fossils of the early *Homo* at Koobi Fora in Kenya. A beautiful skull, with the unimaginative label of KNM-ER 1470, showed a wonderful array of primitive and advanced features, and for many scholars confirmed the existence of *Homo habilis*. This dampened the controversy for some time. It was clear to all scientists studying the material that some kind of hominid with a larger brain yet primitive facial features existed between *Australopithecus africanus* and *Homo erectus*. But the eventual status of 1470 and other early members of *Homo* was to become much more imaginative.

Name Droppers

Now fasten your seat belt and let's take a ride through taxonomic hell as we briefly quibble about names. Most scholars in this peculiar branch of anthropology now accept the legitimacy of *Homo habilis*, but almost all disagree on which fossils represent the species. Some of the fossils, including KNM-ER 1470, have now been referred to a species called *Homo rudolfensis*. Others are best seen, by some taxonomists, as *Homo ergaster*, an early African version of Asian *Homo erectus*. A single fossil hominid from the South African site of Swartkrans enjoyed the long since forgotten name of *Telanthropus capensis* before it became *Homo erectus*; it has since been assigned to either *Homo habilis* or *Homo ergaster*. (Oddly enough, half of the fossil had originally been referred to *Paranthropus robustus* before being found to be the indisputable part of another fossil fragment representing *Telanthropus*.[6])

Hold on, it gets worse. Some putative *Homo ergaster* fossils may be attributed to *Homo erectus*; some *Homo erectus* fossils may be *Homo rudolfensis*, whereas others may be *Homo habilis* or *Homo ergaster*. Confused? Even to a professional it is quite perplexing if not disturbing, and this is just a sampling of the taxonomic options.[7] Hominid taxonomy is in a state of chaos, in the nonmathematical sense.

Thomas Huxley commented in 1863 that scientists "cannot settle to this day which is a variety and which is a species."[8] The same is true now, with both fossil and living animals, and will be true for the foreseeable future. Thus I find the debates about early hominid taxonomy to be particularly specious, if you will forgive the forced pun. Darwin taught us long ago that species were not fixed and immutable entities, and indeed he found the use of the term to be "vague and arbitrary."[9]

Taxonomy may be a necessary evil, for we need effective terms of reference if we are to discuss these fossils. But the arguments about species status are incessant and perhaps unresolvable. So my attitude is that if we cannot agree upon an answer now and may not ever agree upon an answer, then we must be asking the wrong questions. Typological taxonomy is a human idea imposed upon nature and must not detract from the lessons inherent in the living and fossil worlds about variability within and between evolving lineages. Once we get beyond the name game, the fossils can provide many lessons—if we ask the right questions.

So what have we learned? I believe that the answer is quite simple: sometime around 2.5 million years ago there appears to have been a considerable increase in the diversity of coeval hominids (fig. 4.1).

Robust Australopithecines   Early *Homo*

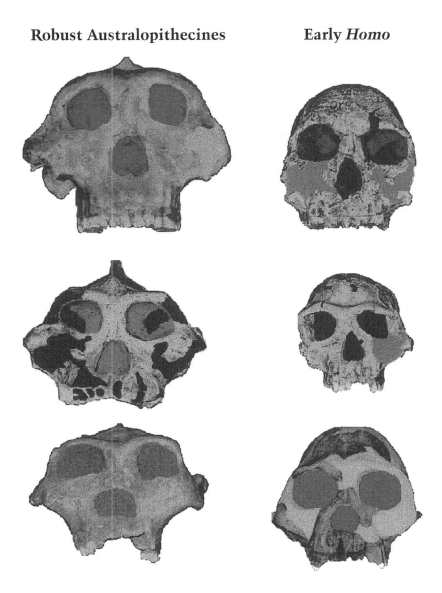

Figure 4.1 ◆ Faces of the great hominid divergence, with some robust
australopithecines on the left and a sample of early *Homo*
on the right. The top four skulls are from East Africa, the
bottom two from South Africa.

Certainly there was a divergence between large-faced, small-brained forms and small-faced, large-brained varieties. That divergence is important, for the robust types eventually went terminally extinct whereas the *Homo* lineage(s) proliferated and led to you and me. The hominid divergence, be it of two species or more, appears to have been a punctuated event setting human evolution on a new course.

There is much to be learned from the divergence of at least two lineages of early hominid, the "robust" and the "gracile." Our ancestors capitalized on the increased variability accompanying their growing populations and apparently took the advice of Yogi Berra: "When you come to a fork in the road, take it." Even among the gracile, those which I call early *Homo,* there is considerable diversity associated with the expansion of the brain and the first appearance of stone tools. The morphological diversity and cultural initiative is interesting in and of itself, without our trying to impose Linnaean categories upon it. And there are many valuable questions we can ask of the fossils. For example, what suddenly changed the early australopithecines into novel forms? Can we find and test evidence as to the causes of our origins?

The Long and Winding Road

The path of human evolution has been full of interesting twists and turns. The origin of bipedalism must have been quite a thing in its day, but our fossil record is still a bit sketchy on the details of the momentous time when our ancestors began to stride into humanity. No less important is the origin of the genus *Homo,* for the advent of the genus was when our brain expansion accelerated and we started to make and utilize stone tools.

When *Homo* arose on the African continent around 2.5 million years ago, it was a time of change for other mammals as well. New types of bovids, the antelope and buffalo that Africans know so well, were beginning to roam the growing savanna. Other animals, particularly some of the large carnivorous beasts that may have menaced poor *Australopithecus,* began to disappear from the landscape. Indeed the earth itself was changing: cooler temperatures arrived with the onset of the modern ice age, and the African continent changed as well. As if to emphasize the dynamism of the time, the same tectonic forces that cause earthquakes today were raising the eastern half of Africa to unprecedented heights. What a challenging time it must have been as the new world order arose.

The apparent coincidence of many changes occurring around 2.5 million years ago has excited a number of imaginations, most notably that of Elisabeth Vrba. Vrba was working at the Transvaal Museum, in the heart of South African fossil country, when she first noted that distinct changes in the African bovid fossils she was studying corresponded with changes in other taxa. Primitive versions of buck were giving way to new and varied grazing antelope. Following on some of Bob Brain's insights, she proposed that environmental change, triggered by global climatic change and the lifting of the African continent, actually *caused* the evolutionary events among African mammal species.[10] This dramatic wave of change included the origin of the genus *Homo* around 2.5 million years ago. It was not a lone coincidence, for a later climatic event around 900,000 years ago preceded the start of another pulse of change in the fauna, perhaps setting the stage for the earliest *Homo sapiens.* A pattern of cause and effect seemed to have been established.

Vrba, at Yale University as of this writing, refers to her evolutionary concept as the turnover-pulse hypothesis: major changes in the mammalian landscape take place in simultaneous pulses when forced by environmental disruption. This notion in many ways capitalizes on the punctuated equilibrium theory. But whereas punctuated equilibrium strictly refers to speciation, *turnover* can include migrations into and out of an area as well. Vrba's concept is a bit fuzzier, for the sudden appearance of an animal in the fossil record can be attributed to movement from someplace else rather than to evolutionary novelty alone. Nevertheless there is no doubt that paleontologists must consider emigrations and immigrations as well as speciations and extinctions. Vrba must be credited for focusing paleontologists' attention on species movements as well as their evolutionary origins.

The turnover-pulse hypothesis explains turnover, and with it speciation, in terms of environmental causes. Speciation, extinction, immigration, and emigration are all seen to be directly affected by climatic change and geological disturbances. The hypothesis implies that not much else happens without environmental change—that species tend to remain at equilibrium with their environments. Thus, with a kick from the fluctuations of global and continental climate, all these changes among mammals should occur at once. In other words, if we look closely, we may find that all sorts of evolutionary and ecological events coincide.

Why should climatic change have caused the origin of *Homo*? There is no shortage of ideas. Most popularized notions shift the savanna hypothesis from the origin of bipedalism to the origin of *Homo*. For example Stephen Stanley, in *Children of the Ice Age*, sees australopithecines as semiarboreal creatures who lived in trees part time and were forced by shrinking forests into complete terrestrial adaptations.[11] Through a complex of developmental and morphological shifts, this allowed the brain to expand to the new levels seen among *Homo*. In *Eco Homo*, Noel Boaz attributes the changes in the hominid brain to the struggle for survival on the savanna—a dangerous place indeed.[12] There is no doubt that the behavioral flexibility and adaptability that comes with a large brain would have afforded great survival advantages. But we already established that *Australopithecus* lived in the savanna as well as the forest. Why did the brain expansion and elaboration not evolve earlier? As the early hominids traversed the sheltered forests of Makapan and Maka as well as the open, arid lands of Taung and Laetoli, would not a bit of cunning have been advantageous? Once again, the savanna hypothesis does not provide a fully satisfactory explanation.

Vrba's hypothesis is a bit more sophisticated than the old savanna origins hypothesis. She sees environmental change as catalyzing the evolution of species for reasons other than adaptation to new environments. Her notion is based on the assumption that speciation events—the origins of new species—occur in small, isolated populations. The theoretical backing for this assumption comes from certain factional ideas about population evolution that emerged from the synthesis of genetics and evolutionary theory in the 1920s. In some scenarios, the process of change in small populations depends on "genetic drift," which we shall investigate more fully in a later chapter. For our purposes at this stage, genetic drift can be viewed as a change in the genetic makeup of small populations due to chance. In small populations these chance changes can accrue quickly, distinguishing one such population from another.

In a nutshell, Vrba proposes that environmental change tends to fragment the habitats of species. For example, if the forest declines and the savanna spreads, then those species living in the shrinking remnants of forest become isolated from their erstwhile mates in other patches of forest. Animals would then certainly emigrate and immigrate to their preferred habitats, hence the geographic component of "turnover." Furthermore, such a situation, through genetic drift, provides a patchwork of small populations ripe for evolutionary change and speciation—a very important part of turnover.

In many respects the turnover-pulse hypothesis is not a bad scientific idea, for it is testable. Likewise any notions of *Homo* origins through ecological change, such as those of Stanley and Boaz, can be tested, at least in part. If we test these notions against the fossil record, we should see coincidences of speciation with events of environmental change. But we must be careful not to see only what we want to see in the fossil data.

Coincidence

Where we look for coincidence, we find it. Indeed, there is little doubt that *Homo* appears as other significant changes occur in the ecological landscape. But one must be wary of interpreting cause and effect from coincidence, as with the coincidental births of Charles Darwin and Abraham Lincoln. Climatic correlations are all too easily made. It was not that long ago in South Africa that miniskirts were blamed for a particularly insidious drought. This conclusion was drawn by supposedly well educated people. I don't know whether they reckoned it had to do with increased reflectivity from female thighs affecting the local weather or suspected that the rain gods simply did not approve of women's knees. Either way, the idea was simply ludicrous. The correlation between skirt length and rainfall was, on the other hand, fairly convincing.

The problem of discerning the difference between mere coincidence and true causation, especially in evolution as represented by an incomplete fossil record, constantly intrigues paleontologists. Could it have been climatic change that propelled the origin of our genus *Homo*? A careful look at additional fossils soon shed new light (as well as some confusing shadows) on Vrba's bold and admirable hypothesis.

Unfortunately for her hypothesis, the evidence for a rapid, widespread evolution among African mammals began to fall apart once the studies went beyond Vrba's bovids. The contrary findings started with Laura Bishop's study of suids—the family of warthogs, bushpigs, and their swinelike ancestors.[13] We have a good fossil record of suid species, but in her Ph.D. work at Yale Bishop could find no dramatic pulses of change in them or in their ecological adaptations. They did not fit Vrba's model. Moreover, Bishop suggested that there were a variety of habitats our hominid ancestors shared with the pigs—a quick shift to the savanna was not in evidence.

Close inspection of a broader range of animals also failed to reveal pulses. A team from the Smithsonian Institution, led by Anna K. Behrensmeyer, studied a complete set of fossil mammals from the Turkana Basin of Kenya and Ethiopia.[14] This is an area where many hominid fossils have been recovered along with a wealth of other animal fossils; it represents one of the most complete fossil records known. At first glance Vrba's pulses appeared to be in evidence. Yet the gaps in the record, and the biases that emerged from small samples at many sites, proved to be deceptive. Once the biases of the fossil sites were accounted for, the apparent pulses were washed away. Change did occur throughout the sequence, and ecological change was in evidence that would have affected the evolution of our hominid lineage. It just did not all happen at once.

So why did Vrba see pulses among her bovids, whereas others did not among other mammals? What can we learn from such an incomplete and biased fossil record? Is the apparent coincidence between climate change and the advent of new hominids and bovids a *mere* coincidence, or might it be important after all? Computer simulations, our closest approximation to a paleontological Rosetta stone, help to fill in the missing pieces of this evolutionary puzzle. They bridge the gap between the fossil rocks and the theoretical hard place.

Model Building

My favorite way of contemplating evolution, and indeed one way in which I derive the most recreational pleasure in general, is by observing the wild animals of nature as we see them today and speculating about their origins. This is the job of an evolutionary theorist. The attraction of working at remote places such as the desert of Taung and the forest of Makapansgat lies not only in the precious fossils that such places hold but in the diversity of life still exhibited. Observations of life today lead to countless mental journeys into past environments and suggest how they may have shaped our uniquely human path of evolution.

The contrast between the environments of Taung and Makapansgat today parallels the variety of environments to which *Australopithecus* was well suited. Even at that early stage, our bipedal gait and precariously balanced heads atop the shoulders were quite sufficient to get us across the dry grasslands of ancient Taung as well as the dark, wet forests of prehistoric Makapansgat. Today I feel equally comfortable, in fact quite happy, in both environments, as does the so-called savanna monkey. So if the environment changed in the past from a dominance of forest to a preponderance of grasslands, what possible evolutionary effect could it have had on our ancestors? Perhaps by observing nature as it is today, for us and other living beings, we can establish some principles to apply to the past.

The problem with observing nature today to test notions of past evolution is that life evolves too slowly. Evolution does not normally unfold in front of our eyes, although some interesting contemporary examples do exist in living forms ranging from bacteria to birds. But for the most part, inspiring though nature may be to watch, it cannot be rushed.

Fortunately, my love of nature is sometimes rivaled by the joy I derive from tinkering with computer programs. By merging the principles of nature with the phenomenal speed of computers, the leisurely pace of evolution can be substantially quickened. Each year of human evolution can be re-created not in one long second, as in our mental exercise, but in a matter of nanoseconds. The entire course of human evolution can be simulated on a computer in a *truly* punctuated event lasting only a few moments while I sip my coffee.

It is quite a challenge to reduce the wonders of evolution over millions of years into a highly condensed computer version that can literally flash before one's eyes. But there is almost as much pleasure to be derived from electronically re-creating nature as there is from absorbing nature's wonders in the wild. Moreover, by varying the computer program's parameters, one can re-create nature in a thousand different ways in order to test the effects of different evolutionary components. In these exercises of the mind, which I find so enjoyable, the meanings of re-creation and recreation merge.

In order to study human evolution on the computer, and test the results of a simulation against what we do know of the past, there are two necessary steps. First, the evolutionary process itself must be simulated using a hypothetical model. For example, we can try to evolve a community of mammals under a turnover-pulse model, or the mammals can be evolved assuming a gradual, constant rate of speciation.

Most evolutionary theorists would stop right there. One can observe tremendous wonders about due process of evolutionary law with such models. But, oddly enough, many evolutionary theorists often do not want to take the procedure further and compare it to the fossil record, which is much too incomplete and messy for their mechanistic theorizing. Even more oddly, most paleontologists do not want to hear about mechanistic evolutionary theories, and prefer to interpret their fossils in what is nearly a theoretical void. To a fossil maven, evolutionary theory is a grin without a cat, as the old saying goes. To an armchair evolutionary theorist, the fossil hunter's endeavor is mere stamp collecting. But an unholy union between the two is really unavoidable if we want to make theoretical progress and test evolutionary hypotheses with real data. It is thus time to bring evolutionary theorists to the high table of human evolutionary studies.[15]

A simulation of evolutionary process alone would, in theory, contain no gaps. It could not be compared directly to what we know, for our knowledge of the past is limited to the notoriously incomplete fossil record. Just as paleontological sites must be viewed in many different lights—from the midday sun to the paleness of the moon and stars—the fossil records derived from them must be viewed in different intellectual lights. Hence the simulation of evolution must be taken a step further in order to mimic not only the process of change among animals but also the chancy formation of a fossil record. The simulation must shed light, that is, on what *fossil samples* we should expect to glean from the far richer trove of evolving animals.

We cannot assume, in our modeling, that all the mammals existing at a certain time become fossilized, or even if they do that paleontologists will find them. After seven years of digging at Taung and not finding any early human remains, and no better luck at Makapansgat, no one knows more about the incompleteness of the fossil record than I. This must be taken into account.

Once a computer program has simulated evolutionary processes among mammals as well as the nature of incomplete preservation in the fossil record, only then can the results be compared with the few fossils in the actual record that we wish to understand.

Creation

The first task of the computer simulation is to create the species we intend to evolve in an electronic world.[16] Normally it is not ethical in science to fabricate data, but in simulation this is a goal. Like real species, the simulated creatures will occupy a space (albeit in a computer memory) for a particular length of time. Although represented by numbers, we can imagine them as *Australopithecus africanus* or *Makapania broomi*. The creation, evolution, and extinction of species is controlled by the parameters we set.

The initial number of species in the simulated animal community or the rate of turnover are examples of the parameters of the computer model. We want to build a model around known or suspected details. Some parameters are exact; some are educated guesses, based upon the original African data. All the model's parameters are derived from the empirical data—observations gleaned from the fossil record or from life today. In this way we can scale the model to fit what we know and then cut it loose to simulate what we don't know.

Now the fun begins. We create an imaginary community on the computer, then set it to life. Animals start to evolve, or at least a turnover of species ensues. The simulation proceeds at 100,000-year intervals, which is about as high a resolution as we have in the fossil record anyway. At each interval some species meet their demise while others evolve into new species. For example, if the given rate of turnover is four species per 100,000 years, then four of the previously existing species go extinct and four novel species make their debut on the African landscape. In the next interval the same happens again. There's no waiting for this drawn-out process to take its time; it all happens in an effortless instant of computer electronics.

The trick in this simulation is to determine which species go extinct and which ones evolve. One could build very complicated, ecologically driven models to determine the fate of species, and eventually somebody may give it a try. But that would involve more parameters, more assumptions, and more complications than are necessary simply to test the hypothesis about turnover pulses. Instead we can rely on an old friend that requires few assumptions: chance.

Chance is easy to program into a computer simulation, for most programming languages have random number generators. So of our species present at any one time, the computer can choose four at random to go extinct. Perhaps species number 27 gets snuffed, or it may live to tell the tale for another 100,000-year period, surviving until the capricious nature of the computer's random number generator sniffs it out for extinction. No one species in such a simulation is given any advantages or disadvantages; all are created equal. Some survive for a mere 100,000 years before the Fates call a halt to their existence, while others calmly persist for a million years or more. The simulation model is totally indiscriminate when it comes to creating beings or removing them from the face of the earth. Such is life in our computer world—and possibly in the real one as well.

The computer program continues to chalk up origins and extinctions of mammal species until the present, thus creating a database comparable to the real world of evolution and turnover. But the real world of past times is largely lost to our observation, excepting the meager but significant offerings of the fossil record. So the next step is to simulate fossilization in order to create a reckoning of what we might find in our excavations across Africa.

Up to this point, evolutionary origins and extinctions have been simulated largely by chance, but within the constraints of parameter values we know or suspect to be reasonably good approximations of reality, such as the total number of mammal species existing at any one time and average turnover rates. Reasonable parameters of fossilization can also be determined by our knowledge of the fossil record, and the unknowns can again be simulated by using our friend chance as a tool. For example, we have identified 55 species of mammals from the oldest *Australopithecus africanus* deposit of Makapansgat. Those deposits are dated between 3.2 and 3.0 million years ago. So why not choose from the simulated mammals available within that time period a random selection of 55 species for fossilization? Likewise at Taung, a bit later, why not choose a random sample of 33 species to represent the 33 we have found in the Hrdlička deposits, to create a simulated fossil record from the evolved species available at that time? And so on for all the known fossil sites in a region, thus scaling the computer model to our particular fossil record.

Why not? Because we know very specific taphonomic events that were happening at those sites. Predators with specific tastes in food were bringing carcasses of their favorite delicacies into the caves from which we subsequently excavated fossils, so perhaps the real fossil sample was not truly random. But, as we do not now know the preferences of saber-toothed cats or extinct hyenas and porcupines, their specific impact on the fossil record is difficult to model accurately. This is especially true since more than one agent may have been responsible for the original deposition of bones in the cave. So, again, chance can be used to simulate specific events of fossilization in the past.

At this point the computer is not the least bit weary, for it has had very little to do. All it has been instructed to do is create a community of species at random, evolve them randomly at a given rate, and then randomly deposit a specific number of species samples at known fossil sites of known ages. The procedure is that simple. And it primarily involves randomness heaped upon chance, albeit scaled to known parameters.

A single simulation, run on a typical personal computer, takes about as much time as it takes me to sip one swallow of the coffee that keeps me awake at night while I create these digital worlds. But given the amount of randomness and chance involved, a different result emerges each time I run the simulation. Such is the nature of chance, just as if by chance the first-born child of my parents had been a girl, satisfying my mother's desire for a daughter, in which case I would never have been conceived. But if my parent's saga were run a thousand times over, within the intimacies of a computer's silicon chips, all sorts of things could happen. So I ran the African simulations again and again, one thousand times to be specific. This took less time than I needed to fix the next cup of coffee. But no coffee was necessary to keep me awake when I first saw the results of simulated evolution and fossil deposition in South Africa. Indeed, my computer spewed out data that left me wide-eyed and made me shiver.

Imagine

I know that a computer simulation is really nothing more than an elaborate form of number crunching, but each time I run a series of simulations I still envision herds of zebras galloping across the silicon plains of my computer with digital carnivores hot in pursuit. Life's rich pageant of evolution plays itself out, over and over again, each time with different animal species surviving the Darwinian struggle for computerized existence while others are condemned to be erased from memory by terminal extinction. But real lessons emerge from the conjuring of imagination.

We know that species in our computer world are arising and going extinct at a constant rate, because that was a simplifying assumption of the model design. The amount of species biodiversity did not change with constant turnover. Perhaps an antiquated zebra was lost, but an antelope was gained. Thus, while the players on the silicon field would change, their numbers would remain the same. If one were to make a graph of the number of origins and the number of extinctions in each 100,000-year period across 3.2 million years, the graph would be a simple straight line. The pace of evolution would be predictable and uneventful—no pulses.

A graph of species turnover becomes considerably more intriguing once the nature of the fossil record is imposed upon the steady state of evolutionary change. We can no longer look at the origins and extinctions of the animals themselves but only at their first and last appearances in the fossil record. In other words, although an electronic animal may exist, it is not necessarily dragged into a cave by our voracious electronic predator who is picking animals at random. Thus a mammal species could be present for hundreds of thousands of years before it makes its first appearance in one of the cave sites that holds the fossil record. Perhaps it was gallivanting around southern Africa when Makapansgat was filling with bones but was not caught in the curious circumstances of fossilization until it wandered past the caves at Taung and later at Sterkfontein. Likewise, a species may have been around long after Taung but somehow may have missed getting dragged into the Swartkrans cave along with *Australopithecus robustus.* So the fossil record may indicate the first appearance of the species at Taung and the last appearance at Sterkfontein, even though its origin was earlier and its extinction later.

The graph of first and last *appearances* is thus likely to be bumpy, as opposed to the straight-line graph representing steady species turnover. But how bumpy? One thousand simulations of evolving species in southern Africa can produce a lot of varying data, given the randomness of our fossilization process. The simplest way to treat the data is to look at the average number of new species that appear in the simulated fossil record for each time period, which gives us an idea of the kinds of patterns to expect. For each simulation run, this fluctuating number of first appearances can be plotted on a graph and compared with actual numbers gleaned from the real fossil record. As a way to summarize the results from one thousand simulations, the maximum and minimum simulated numbers of first appearances can also be plotted for each time period, so that we know the range of possibilities simulated. Such a tidy representation of the simulation results can be found in the accompanying graphs (fig. 4.2).

In the first graph you can see the trends in data from the actual fossil record we have of South Africa. The pattern has apparent peaks and dips in evolutionary activity (first appearances of species in the fossil record, in this case). The second graph shows the *simulated* patterns of a bumpy fossil record from the computer model of constant, unwavering species turnover. There is an almost frightening correspondence between the averages for simulated first appearances and the actual first appearances of species we see in the real fossil record. These averages are accompanied by an envelope of simulated possibilities—maxima and minima from one thousand different scenarios, all showing the same pattern of peaks and valleys. Time after time, the simulated fossil records clearly show the same *apparent* turnover pulses. But how could that be, when the simulated rate of evolution was not punctuated or pulsed? Something was amiss in the fossil record.

Despite the constant rate of simulated species turnover, *despite* the amount of randomness used in evolving different species, and *despite* the chance nature of choosing species for fossilization, the average pattern of simulated first appearances corresponded almost precisely to the real fossil data.[17] Sometimes it still makes me shiver.

So what does all this mean? Quite simply, it means that the peak of fossil species first appearing around 2.5 million years ago needs no grand explanation. Nor does the peak some 700,000 years ago, although both peaks occur near times for which important climatic changes have been postulated on the basis of independent evidence. Such peaks, which appear to be pulses of mammalian species turnover, and perhaps significant punctuated evolutionary events, are explicable simply by the wanton nature of an incomplete fossil record. Fossils unfairly distributed at sites spread unevenly through time give the *appearance* of spurts of new species when in fact the rate was steady.

That these deceptive features correspond with supposedly dramatic shifts in climate is a mere coincidence of no particular importance. Like the Makapansgat bones that seemed to be *Australopithecus* tools, and like those which appeared to be burnt, the fossil record fooled us.

Observed First Appearances

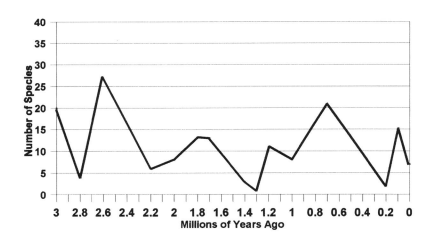

Simulated First Appearances

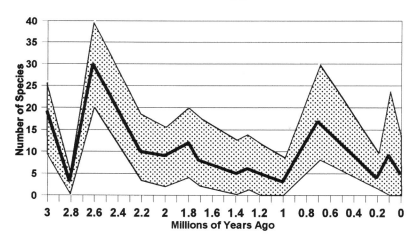

Figure 4.2 • A graph of species' first appearances in the fossil record shows an apparent turnover pattern, but a simulation under conditions of constant turnover produces a similar pattern. *Shadowed area:* range of simulation results. *Dark line:* average of all simulations.

Nevertheless, I continued to examine the simulation results in other lights. Their meaning may be ambiguous after all. With or without the climatic events, these simulations suggest that evolution could have proceeded at its own leisurely, constant pace and still produced the "trends" we see in the fossil record. Seen in this light, the pulsed trends are more apparent than real, and we may be well advised to look for other explanations of the fossil record than changes of climate causing pulses in the rate of speciation. Or—and we always must admit other possibilities—the fossil record of southern Africa just may not be refined enough to pick up pulses of species turnover. Maybe there were pulses, but the incomplete nature of the fossil record missed them.

East Africa provides one test of how good the fossil record may be for discerning changes in the pace of evolution. My first simulations of East African turnover, as in the study of the actual fossil record conducted by Behrensmeyer and her Smithsonian colleagues, produced results similar to those for southern Africa: random speciation at a constant rate, and random fossilization at East African sites from Ethiopia to Tanzania, can produce the pattern of peaks and valleys in species appearances that we know from the fossil record.

Can turnover pulses do the same thing? That possibility is easily tested: all we need to do is run another thousand computer simulations, this time programming them to include pulses of origins and extinctions at the times of the supposed global climatic events. If such turnover pulses actually occurred and the fossil record is just too weak to pick them up, then the new simulations should fit the observed data just as well as the first simulations involving constant, steady, mundanely regular turnover.

It was with some trepidation that I went back to the drawing board and redesigned my computer simulations to include turnover pulses. After all, my simpler model of constant turnover and chance fossilization had generated such pleasing results (although Elisabeth Vrba and other proponents of the turnover-pulse hypothesis were somewhat less charmed).[18] But good science involves continuous testing and retesting—good theories and bad all go through a process not dissimilar to natural selection. Only the fittest theories survive the selective onslaught of scientific scrutiny. Could the turnover-pulse hypothesis measure up to the empirical data?

That question was easy to test just by increasing the turnover rate at specific times, such as that leading up to the supposedly important climatic event of 2.5 million years ago. For the remaining time periods, with no dramatic changes in climate, the pace could be slowed to more of an equilibrium involving few speciations. Initially, with small pulses, the new simulation results were not much different from before. The simulated fossil records were not good enough to detect the pulses of speciation built into the program. On the other hand, with such small pulses, constant turnover still accounted for most of the species origins and extinctions, as well as first and last appearances in the fossil record. Perhaps a bit more stringency would put the turnover-pulse hypothesis to the test.

One more try. If the turnover-pulse hypothesis were of primary importance in explaining the real world of evolution, then one would think that turnovers of African mammals would most often occur in pulses. Pulses would be the rule rather than the exception. Evolutionary origins and extinctions, as well as immigrations and emigrations, should tend to occur more frequently during periods of dramatic climatic flux than at any other time. Beethoven should be more influential than Mozart. The rest of the time stasis should ensue, with the same musical themes played over and over again by the same species. So just for fun, why not simulate pulses as the dominant mode of turnover and let the intervening periods contain minimal turnover (and unimportant evolutionary compositions).

Unfortunately for the turnover-pulse hypothesis, at this point the simulations began to stray from the real African fossil records. As the Fates would have it, the strong turnover pulses I had programmed into the model could now be observed in the simulated fossil records (just as such pulses appear to exist in the real fossil record), but they still fell within the realm of simulated chance. Chance is a sneaky operator. On the other hand, the observed fossil records of some intervening periods between pulses begin to fall outside the window of simulated chance. So the pulses may or may not have been real, but the stases in between looked dubious.

What can we conclude from this interesting but somewhat arduous exercise of computer chips? First, we can say that we have insufficient evidence of evolutionary pulses (or turnover pulses of any kind). We cannot eliminate the possibility of small pulses, perhaps occurring in concert with climatic change, but the chance nature of the fossil record is at least an equally feasible explanation of what we observe. Large pulses are another matter. The seemingly concurrent appearances of large numbers of species are no more significant than the coincidental appearances of famous American statesmen with eminent scientists across the Atlantic; they are nothing more than the deceptions of an incomplete record of the past combined with a wishful look for meaningful trends. The equivocal evidence opens wide the door to other explanations of the pace of evolution and the appearance of early *Homo*.

Those who wish to hold to the turnover-pulse hypothesis may gain some solace from the lamentations of no less than Charles Darwin, who asked in print: "Why does not every collection of fossil remains afford plain evidence of the gradation and mutation of the forms of life? We meet with no such evidence, and this is the most obvious and plausible of the many objections which may be urged against my theory."[19] Thomas Huxley jumped to Darwin's defense, after his "candid avowal": "What he says in effect is, not that paleontological evidence is against him, but that it is not distinctly in his favor; and, without attempting to attenuate the fact, he accounts for it by the scantiness and the imperfection of that evidence."[20] The fossil evidence *did* eventually come to confirm Darwin's hypothesis, and it may still justify the turnover-pulse hypothesis—in part.

The more important conclusion is this: if there *are* pulses of evolution, then they are not any more important than constant turnover of species in both the digital and real worlds. Pulses may or may not need a cogent explanation, but constant turnover in the absence of climatic causes *certainly* needs a good evolutionary explanation. The best evolutionary theory would be able to account for both pulsed *and* constant turnover; it would allow for punctuated equilibrium as well as Darwinian gradualism; we could appreciate both Beethoven and Mozart.

On the Origin of Species

In our search for explanations of human origins, surely we can assume that environmental change, be it change in the climate or change in plant and animal communities, must *cause* change in evolving animals. Or can we? Environmental change was a large part of the Darwinian vision, was it not? Well, no, not exactly. We need to clarify a point or two.

Darwin, like Lamarck before him, initially felt that environmental change was terribly important for the initiation of new variants and new species. This was largely due to his retrospective observations of island species, particularly the birds and tortoises of the Galápagos Islands. These varied species seemed to have adapted to their environments upon separation from others of their ancestral stocks. In his unpublished essay of 1844 on evolution, Darwin envisioned periodic geological uplift and subsidence of an earth in flux as the catalyst of new species, with isolating mechanisms being the key. But later, with his study of barnacles—spread across a relatively constant sea—he realized that the variability upon which natural selection acts needs no catalyst. Barnacles, he saw, have evolved a tremendous amount of variability without any outside force to prod them along. It was an important change in Darwin's thinking when he realized that no environmental kick was necessary.[21] Things varied on their own, and natural selection worked from that intrinsic starting point.

By the time Darwin made his theory public, he was actually quite adamant that environmental change was *not* necessary for the evolution of new species. At least in the first edition of *On the Origin of Species by Means of Natural Selection*, published in 1859, he dealt with the problem at length in no uncertain terms. In later editions he softened his approach but never abandoned the principle. "Nor do I believe," Darwin assured us, "that any great physical change, as of climate, or that any unusual degree of isolation to check immigration, is actually necessary to produce new and unoccupied places for natural selection to fill up by modifying and improving some of the varying inhabitants."[22] His point was illustrated by the many animals that live in a wide variety of environments and by plants that could be moved quite successfully from around the world to new environmental settings in England. Darwin of course noted that natural selection would ensure that the best-suited among the existing variants of a species would dominate in a particular area, but he was not particularly concerned by moderate modifications of that environment.

So what *does* happen when the environment changes? Evolutionary adaptation is a possibility; but so are migration and extinction, as noted by Vrba. At this point it may help to move into the evolutionary theorists' realm. We can momentarily set aside the fossil data, incomplete and fragmentary as they are, and move into the observable world of nature evolving. There we will find and test, among living creatures, the principles of environmental change and evolution. These creatures need not be as close to us as my much-favored baboons. After all, Darwin acquired many of his evolutionary ideas from studying barnacles. Darwin's home was cluttered with the barnacles he had collected. Huxley had derived many of his ideas from studies of the jellyfish he found at sea; his shipmates aboard the *Rattlesnake* had to endure the scattered bits of jellyfish anatomy from Huxley's assiduous dissections. No, one need not look hard to find the marvels of evolution, for they appear in even the lowliest of creatures.

Indeed, one need look no further than one's kitchen to see evolutionary principles in action. You may not want to admit that the Darwinian struggle for existence is going on in your own home. Unashamedly, although with a degree of embarrassment, I'll admit it. In my own kitchens of former homes, as far apart as Missouri and South Africa, breeds of cockroach fought for survival. In the process, these noble insects provided lessons that are equally available to any observer of nature, including lessons about turnover pulses and the limits of natural selection in changing environments.

Darwin had his barnacles. Huxley had his jellyfish. I have my cockroaches.

A Pulse of Roaches

I am sure that my first experience of cockroaches must have come when I was a boy in Ohio. My mother would not like to admit it, for she kept an immaculately clean house despite the presence of four messy boys. But the odd cockroach, and an occasional mouse in the basement, crept into our comfortable suburban home. Life forms of many descriptions are amazingly persistent and adamantly pursue all environments.

But my first real experience with cockroaches was in St. Louis, Missouri, where insect life finds a warm, humid, and happy home. As a graduate student there, like university graduate students almost everywhere, I lived for a while in a tenement home. It was affordable and acceptable to a young aspirant anthropologist. And there, in the heart of an urban environment, insect life thrived. As a student of evolutionary science, I should have been endlessly fascinated by cockroaches in my apartment. Their origins were in the caves of Africa, so it was only natural that they eventually spread into our warm homes.

And from a Darwinian point of view they are a tremendously successful and diverse group of over 3,500 recognized species. Their sheer numbers, with a potential for 400,000 offspring from a single pair within a year, ensure an endless stream of variants on which natural selection can work. Perhaps that explains the large number of species.

However, I was unimpressed by the occasional cockroach that strode through my apartment. Despite my reverence for all sorts of life, the dispassionate sole of my shoe put a quick end to the odd cockroach that waltzed across my floor. Natural selection had begun. I was totally nonplussed by the invasion of cockroaches that came with the painting of the apartment next door. It was a turnover pulse beyond proportion.

Late one night I woke from thirst, as one often does in the heat of a St. Louis night. I went to the kitchen of my humble home, which was adjacent to the newly painted apartment being readied to house the next poor student occupant. In the dark, my kitchen table did not look right; it was not as still as a solid wooden tabletop should have been. Bleary-eyed and somewhat apprehensive, I turned on the light. Urghgh! The table was completely covered with a swarm of black cockroaches. In an instant they dispersed and were gone, disappearing amid the cracks that characterized the walls and floor of my very own tenement home. As an agent of natural selection, I had the opportunity to squash only a chance few.

On occasion I allow my passion for nature to wane—this was one such time. But in retrospect it taught me something about turnover pulses. The real world of nature *does* involve changes of climate, changes of environment. But evolutionary adaptation to a new environment is only one option, and an outside option at that. Migration or extinction are the other options.

Before the apartment next to mine was painted, many cockroaches found it a friendly, cozy home. Paint was not to their liking; it was the equivalent of a change in climate, and certainly a change in atmosphere. Just through the wall, which they had no problem penetrating, was a much more inviting home. And there was food on the table, albeit fine scraps I had not noticed. Whatever the circumstances, my kitchen now provided a new home.

Could it have been much different in the past? When the African climate changed some 2.5 million years ago, what would the bovids have done? Would they be any different from cockroaches in a painted room? Sure, some would stay behind and die. Tough luck, in the eyes of nature. That is strong natural selection. But most would move to a new environment, where the grasses or leaves they enjoyed were still aplenty. Although such migrations undoubtedly occurred, as they do today, they are difficult to pick up in the fossil record—areas as large as southern or eastern Africa contain many habitats for small-scale migrations. Likewise, the repulsive cockroaches were repelled from one kitchen to the next, but the migration was not noticed on the scale of a city such as St. Louis. Immigrations and emigrations are important parts of Vrba's vision of turnover; that much we must accept.

So as a matter of principle, an important question arises. When the climate changes, why should species evolve rather than follow their accustomed habitat? Over the past 3 million years in Africa, temperatures and rainfall have changed from place to place, but most animals could still find their place. An animal, and even a plant, can move a few degrees north or south with the climatic regime, just as cockroaches can move next door when the paint fumes become a bit much. And just as no cockroach evolved to thrive in paint fumes, why should any mammal evolve to new environmental circumstances when its preferred habitat is just next door, perhaps a few miles to the north?

My great-grandmother, who lived in homes across the United States from Ohio to Oregon and back again, was fond of assuring me that "it's cheaper to move than to pay rent." One can be sure that the same applied to my more distant ancestors in Africa.

On the other hand, you can run but you can't hide. Other animals, like other people, will also move when the going gets too rough. Things might get a bit crowded in the promised land to which they flee, and competition to establish a home would ensue among various species. "Out of the frying pan and into the fire," as Great-Grandma said on occasion. With increased pressure from competition, shouldn't these animals be forced to adapt?

Cockroaches, who were lying in wait for my arrival in South Africa, have something to teach us about adaptation as well. For some time they competed with me for dominance of my kitchen in Johannesburg. Despite my attempts to the contrary, it was an inviting environment for the filthy little chitinous beasts. Initially, their influx into my kitchen was controlled by flattening them with any available object, quickly and accurately placed by my adroit hominid hand and guided by my superior hominid vision.

Some cockroaches were quicker than me, and the quickest got away into the confines of tiny places I could not reach. But clearly I was selecting for the fastest and most easily hidden bugs—that is, until I penetrated their hiding places with an insecticide spray. It was a horrible massacre. Many died. Even more, I am sure, retreated to the kitchen of the adjoining apartment to regroup. But some of these little critters were rather impervious to my poison. And their numbers grew.

Certainly natural selection, or perhaps artificial selection, had occurred. The cockroaches had *apparently* evolved a resistance to bug spray. It was time to move to the next phase of our arms race. Maybe in time I could be instrumental in evolving an entirely new species of cockroach. But it was not to be.

Following the strategic use of insecticide spray I noticed that the insidious invaders of my home were a bit slower than before. It was back to easy manual extermination. Perhaps the cockroaches had bred with the bugs next door, who had not been challenged by my ever quickening hands and feet. Or, it could be that insecticide-resistant bugs were just slower. But the entire cycle started over again. They became quicker, and my squashing efforts became less effective. I resorted to spray once again, after they had been free of it for a while, and they died in massive droves. The slower ones came back, and the cycle of violence began anew—until the final day of eradication, or extinction if you will.

But what, if any, evolution had occurred? Initially there was selection for faster insects. But they were not created by the environmental change of my arrival to whack them; they already existed as variants. Likewise, the insecticide-resistant cockroaches already existed, but the forces of selection made them more prevalent in the population when I applied the spray. It was indeed evolutionary change, but nothing new had evolved.

Perhaps cockroaches are not the most delightful creatures to choose as an example of evolution in action. But, like humans, cockroaches owe their success to their adaptability in the face of challenge. Indeed most mammals are reasonably adaptable, so a change in their environment does not necessarily mean that they will evolve new features. It just means that they will either move, like my great-grandmother, or that existing variants, like Darwin's variable barnacles, will undergo the selective process for a while.

A final option of turnover is extinction.

The Flip Side of Environmental Change

In the computer simulations discussed earlier, we examined the possibility of species originating in vast pulses. None were detectable from the fossil data of species' first appearances. Evolution at a steady rate, with incomplete fossilization, could have produced every apparent bump or "pulse" in the fossil record. Last appearances, however, tell a different story. Once again, we do not know when species went extinct, only when they last appeared in the fossil record. But sometime following 2.0 million years ago in East Africa there was an apparent decline in species biodiversity. No matter how I adjusted the parameters of the simulation, I could not mimic the observed imbalance of species origins and species extinctions. The real data always had an excess of extinction. Likewise, in the study of the actual Turkana Basin data conducted by Behrensmeyer and her Smithsonian team, species disappeared at a faster rate after about 2 million years ago, even with the researchers taking the incomplete nature of the fossil record into account. Africa undoubtedly began to lose more mammalian species than it gained. Was this loss the product of environmental change?

If one looks beyond the mammals of Africa to other life forms, ranging from mollusks to plants, one usually finds no good correlation between climatic change and extinction. Most species, as we follow them through the fossil record, tend to "weather the storm" of climatic change through migrations, without dramatic evolution or extinction. Some species, it is true, suffered from the reduction of habitat such as the shrinking forests of Africa, and this increased their chances of extinction. Dramatic climatic change may also be implicated in some mass extinctions—in such pulses of turnover, extinction is relatively quick, and recovery of species diversity is slow. Climatic change, however, cannot always be implicated as the cause of extinction.

Climate is just one part of the environment to which a species must adapt. Early hominids had to adapt to the plants and animals that evolved around them as well. The chance of encountering a predator had to be balanced against the opportunity to find food and water. Every other species with which our ancestors interacted was part of the equation of their survival. Likewise, the hominid presence, like that of any other animal, undoubtedly affected the lives of other species. This intricate web of life was a complicated product of many equations—it was a nonlinear, chaotic system.

An interesting computer experiment, conducted by Canadian zoologists Kevin McCann and Peter Yodzis, clearly illustrates the tenuous nature of the chaotic environmental web that shaped the lives of our ancestors.[23] McCann and Yodzis built a computer model of a simple, small food chain of three species. It was similar in some respects to the Lotka-Volterra model (chapter 1), in which the population densities of a simple predator-and-prey community fluctuated in a nonlinear fashion. The McCann-Yodzis model involved no chance events but was simply a set of deterministic equations.

This computer model went through a series of cycles, with populations fluctuating through similar maxima and minima at fairly regular intervals, and the populations continued to fluctuate in repetitive fashion over more than thirty-five hundred cycles. This was not unexpected. Then something remarkable happened. After all this regularity, the population of the top predator suddenly and inexplicably crashed. Having achieved a seeming balance for cycle after cycle, a population went extinct for no apparent reason. There was no outside perturbation, no simulated environmental change.

Such is the nature of chaotic systems. Aptly named "chaotic blue sky catastrophes," such sudden, inexplicable disruptions to seemingly well balanced systems are often found in complicated nonlinear systems. This is important to recognize, especially because once one part of a system fails, even if out of the blue, it may affect other parts of the system. The same may apply to the sudden success of one part of the system.

How might hominids have fit into the ecological equations? Given the vulnerability of complex and chaotic ecological systems, their impact is difficult to pin down. The advent of early *Homo* and their stone tools certainly must have had some effect on the species around them. Initially, the effect may have been small, but later one might expect the impact of our ancestors to have been greater, especially when these slow but clever bipedal animals learned how to hunt with their crude stone implements. With the evolution at about 1.8 million years ago of the next, dare I say, species, *Homo erectus*, tool use and environmental manipulation eventually became greater. Once hominids took control of fire, our ancestors' competitors and prey alike might have found this evolving species to be quite a nuisance indeed. Today *Homo sapiens* is the greatest nuisance on earth, in terms of disrupting the environmental habitats of many species, leading to yet another event of mass extinction. But when did our devastation begin? Did *Homo* play a role in the rise of extinction that happened after 2 million years ago?

Coincidental Extinctions

The southern African evidence provides some tantalizing clues that the environmental disruptiveness which characterizes our species now may have had an early start. From the Sterkfontein cave site, near the fossil deposit where the late excavator Alun Hughes found *Homo habilis*, above the spot where his ashes now rest, one can look across the Blaaubank River valley and see another cave on the opposite hillside. Although this cave, known as Swartkrans, formed in the same dolomitic rock, it did so at a later time than Sterkfontein. The bones that came in through the opening of Swartkrans thus reflect events of later time periods. It was at Swartkrans that Robert Broom found stunning remains of *Australopithecus robustus*. Living and dying side by side with its robust cousin was our ancestor, *Homo erectus*, the successor of *Homo habilis* in our lineage.

Bob Brain, who revealed the natural origin of Dart's supposed ODK tools at Makapansgat, directed a most meticulous dig for twenty-one years at Swartkrans. The excavated cave deposits chronicle a fascinating story spanning at least a half million years. In each of the lower three deposits, or "members," Brain found more fossil evidence of both *Homo erectus* and *Australopithecus robustus*. Thus, it is clear that after our own ancestors took a fork in the road of evolution and became different from *A. robustus*, their paths continued side by side for a considerable time.

The transitions marked at Swartkrans, unearthed and recounted by Brain and his team, are quite telling. The earliest deposit, denoted in geological parlance as Member 1, dates to around 1.8 million years ago. For the first time in the South African fossil record, we see *Homo erectus* (or a close taxonomic variant) along with its robust hominid cousin. But, out of the forty-seven recognized mammal species from the entire deposit, Member 1 holds the latest fossil appearance for at least eight. Among these eight species that subsequently disappear from the fossil record, presumably going extinct forever, are an interesting array of animals. Some of the large menacing cats that presumably fed on hominids for well over a million years subsequently go extinct. Some large hyenas, including the kind that used their teeth to crunch bones into the Makapansgat fragments Dart mistook as tools, also take their final curtain call after *Homo erectus* marches onto the stage. My beloved baboons, once so diverse (and thus prevalent at every fossil site I excavate), also dwindle in number and shrink in species diversity.

All was not lost, for not only the hominids survived. The bovids in fact thrived. There was a truly remarkable increase in the number of bovid species, perhaps unparalleled by any mammalian family anywhere at any time. Every dog may have its day, but every cow seemed to have had a considerably longer and more profound period of success.

In the evolutionary species stakes, big fierce carnivores and lovable baboons lost, while humble grazing bovids and early *Homo* won. On balance, however, the total diversity of mammals began to dwindle both here in South Africa and in East Africa. And it all seems to have begun about 1.8 million years ago.

Alan Walker, a key figure in American paleoanthropology, proposed that the decline in the large carnivores, and the evolution of the sleeker models we know today, was a direct consequence of the evolution of *Homo*.[24] It seems to have been more than a mere coincidence, and more than an artifact of the fossil record, because *Homo* clearly entered the niche of scavenging and perhaps hunting meat. I once opined that *Homo* was responsible for the decline in baboon diversity, postulating that their tragic demise was due to their ineptitude in the face of an emerging *Homo* evolutionary revolution and competition for resources.[25] Such notions of cause and effect, *Homo* origins and the extinction of others, are tempting to make, but dicey.

Again one must be wary of deception. Correlations of fossil appearances and disappearances do not necessarily reflect the realities of origins and extinctions. Furthermore, it is difficult to pin the blame on any one component of an ecological system in complex, chaotic systems, as illustrated by the McCann and Yodzis model. Sometimes no apparent reason is necessary for the extinction of a top predator such as a saber-toothed cat. One would need to find either direct evidence of hominid involvement in the extinction of others, which is nearly impossible, or implicate the hominids by establishing a long-term pattern of effect. Certainly at later times that pattern becomes clear—

humans cause extinctions—but the pattern of the human effect is a bit fuzzy during these early stages.

One thing cannot be denied. Humans were playing an increasingly important role in African ecosystems, and soon would do so in novel webs of life beyond their home continent. It is not beyond the realm of possibility that early *Homo erectus* may have been the metaphorical butterfly that, upon flapping its wings under blue skies, set in motion a chaotic set of events. Such possibilities become more evident as we continue the African story, as best told at Swartkrans.

The apparent losses of some animals were slowly, but only partially, recompensed by the eventual evolution of new ones. Witnessing these changes in the Blaaubank valley were *Homo erectus* and *Australopithecus robustus*. Unlike the animals around them, neither hominid seemed to change very much. *Homo* had changed a bit; some paleoanthropologists make a taxonomic distinction between the *Homo* fossils of Swartkrans Members 1 and 2. But no dramatic changes occurred: their faces, brains, and bodies had the same basic adaptations, and their tool kits remained essentially the same as well.

In Swartkrans Member 2, representing a time period about 200,000 or 300,000 years later than Member 1, fossils of our two hominid lineages appear again. Perhaps they both thrived upon the continually emerging environment, in which the king of the beasts changed from supercarnivores to smaller, more nimble cats while the modest grazing bovids gained much ecological ground. Nevertheless the fossil evidence contradicts such a notion of apparent hominid successes, for they were still being fed upon by the surviving, indeed thriving, carnivores such as the ever successful leopards. At this stage the hominids were still not living in the Swartkrans caverns as romanticized "cavemen" but were simply dying there. Like their australopithecine ancestors, they were the hunted more than the hunters.

By the time bones started to accumulate in Swartkrans Member 3, a deposit following closely on the heels of Member 2, a significantly new part of our hominid history, or prehistory, emerges in blazing glory: the first recorded use of fire. Our ancestors of the time may not have created fire by their own technology, but they certainly captured and controlled it for their own purposes. What a spectacle it must have been for both them and their foes!

It has been noted that bone, in a hot fire, only initially turns black. Later it turns ashen white and always changes dramatically in its chemical and structural configuration. Many bone pieces of Swartkrans Member 3, 270 to be as precise as Bob Brain's particular methods dictate, were undoubtedly burnt at a high temperature.[26] Even as fossils they reveal the same signature characteristics as the bones in one's campfire. Such burnt bones appear throughout level after level of the excavated deposit. Fire was burning bone in the cave; this was far beyond the realm of chance, for embers falling into the cave could not have consistently produced the same results. Certainly the fire was controlled and repeatedly utilized.

The likely culprits, the ones who captured the power of fire, were early *Homo*. Their tools, created from stone as well as bone, littered the same Swartkrans Member 3 deposits. They were truly *cultural* beings, at least from the perspective of paleontologists and archeologists, in that cultural beings leave artifactual material remains. Fashioned stone tools, especially accompanied by the controlled use of fire, are telltale marks of ancestral creatures that are uniquely ours.

Humankind, as we know it, was born of the artificial transformation of nature into tools of survival. Stones became extensions of one's arms and hands; fire became a friend rather than an enemy. Most other animals, through natural selection, acquired their specific adaptations through more biological, nonartificial means. Humans discovered, quite fortuitously I imagine, the adaptive value of environmental manipulation. Morphological adaptation gave way to cognitive adaptability.

Meanwhile our robust-faced family members must have been having a hard time. In the cave deposits of Swartkrans Member 3 we find the last fossil evidence of *Australopithecus robustus*. The road they were thrust upon by the fickle forces of evolution turned out to be a dead end. For about a million years they lived in the presence of *Homo* but never developed the expanded brain of their more sentient cousins.

Despite their adaptations, both *A. robustus* in South Africa and the hyper-robust *A. boisei* in East Africa went extinct about the same time, making their final appearances in the fossil record at about 1.4 million years ago. The face and teeth of these hominids were well adapted for grinding tough, fibrous vegetation, but their dietary specialization did not suffice forever. Supplies of their preferred food sources may have dwindled as new creatures and a drying environment emerged. But they were hominids, and having survived for so long they must have been adaptable.

It is difficult to ascertain why *A. robustus* went extinct. Richard Klein, no stranger to the southern African fossil record, suggested that early *Homo* outcompeted the dim-witted, lumbering *Australopithecus robustus*.[27] This again is difficult to defend. *A. robustus* was not endowed with a particularly large brain, relative to us, but it was surely quite adequate and even substantial when compared to the brains of other contemporary animals. Although we tend to attribute all the stone tools we find to early *Homo*, the fossil evidence of *A. robustus* challenges that notion: the hand bones reveal no morphological incapability for a precision grip. Perhaps *A. robustus* could have made stone tools as well. And, as a biped, *A. robustus* would have been afforded whatever adaptive advantages such a stance provided—such as freed hands.

But competition comes in many forms. There is no smoking gun, but it is intriguing that among the last known fossil remains representing *A. robustus* at Swartkrans Member 3 was one burnt finger bone. The genesis of our lineage just might have involved events analogous to the biblical story of Cain killing Abel.

The Human Butterfly

Chances are that the relationship between *Homo erectus* and *Australopithecus robustus* was more complicated, involving a bit more in the way of nonlinear dynamics. Their possible associations in daily life were many, given our knowledge of the intricacies of ecological relationships. Due to the limits of the fossil record, we may never be able to discern the nature of their relationship, if any. However, one hypothesis illustrates how nonlinear systems may have woven the web of life and death among the hominid lineages.

As noted above, the diversity of carnivores decreased following the emergence of *Homo erectus*. The extinction of *some* carnivores certainly changed the competitive environment for other carnivores (no matter what the climate was doing). Imagine, then, the following scenario, suggested by Richard Klein. It involves relationships more intricate than direct competition between *Homo* and *A. robustus* but is still a plausible sequence of events.[28]

Imagine a particular large cat that consumes a number of mammalian species, among them the occasional hominid. A competitor carnivore has a taste for hominids but needs other species to fulfill its dietary requirements. Carnivores do not live by man alone! The two carnivores find themselves in competition for other mammals and thus limit each other's numbers. As the ecological system evolves, something leads to the extinction of the large cat. With its demise, the other faces less competition. The survivor's population grows and, somewhat more often, attacks a passing, slow, and bipedal vegetarian: *Australopithecus robustus.* In classic Lotka-Volterra fashion, the *A. robustus* population declines. But unlike the seemingly stable outcome in a simplified two-species model of predator and prey, in which each takes turns crashing and rebounding in numbers, this predator just continues to expand its numbers by feeding upon multiple prey. Among those who continue to suffer population loss: *A. robustus.*

In other words, *A. robustus* may have been shielded from extinction because its main predator was in competition with another predator that had overlapping but different tastes. Perhaps, for example, one of the saber-toothed cats was in competition with leopards. Saber-tooths probably went after some of the larger prey that leopards sometimes hunt, such as giraffes or eland, but usually left alone smaller prey such as the hominid. We know that saber-toothed cats eventually went extinct, as indeed some did around the time of the origin of *Homo erectus,* thus allowing the leopard population to grow. Leopards are known today to consume humans, and there is good evidence at Swartkrans that they did in the past: a skull of *A. robustus* with two puncture marks fit precisely by the lower canines of a leopard jaw.[29] Leopards, not *Homo,* may have been the cause of extinction for our late hominid cousins.

Homo, however, may have had an indirect effect in the chaotic system of nature. First we must ask why *Australopithecus robustus* went extinct while *Homo erectus* survived the perils of a growing leopard population. The answer may be fire. Certainly at Swartkrans *Homo* had the ability to ward off predators from the caves. The caves were formerly inhabited by saber-toothed cats. Could *Homo* and fire have stolen the cavernous homes on which the saber-tooths relied? If so, *Homo* could have set off the chain of events that led to the demise of the large cats and the eventual loss of *A. robustus, Homo*'s evolutionary cousin.

The possibilities are endless. Year after year, one millennium after another, life's complex story unfolds. Novel features evolve in animals, new interactions occur among predators and prey, and there is no telling what will happen next. Climate is but one variable, one para-meter of the complicated equation. Rather than being a well-struc-tured Beethoven or Mozart symphony in which everything happens on cue, evolution is more like a jam session of improvisational jazz musicians. Every species, every individual plays its role in the dynamic chaos of life, but jazzes it up in unpredictable ways. Evolution, as it occurred in Africa, was not merely one of a thousand computerized possibilities but one of a nearly infinite range of nature's prospects.

THE PACE of evolution and extinction varies, prompting scientific minds to probe for causal explanations. What speeds up the origin of species or slows down the forces of change? It would seem that large-scale environmental disruption from earth's climatic shifts should cause a flurry of evolutionary activity—plants and animals should scramble to adapt. Yet the fossil record has not yielded clear evidence of any such sudden evolutionary change across many mammals, not even at times of great climatic change. Indeed, either stasis, migration, or extinction were the more common effects of climatic alterations.

The genus *Homo* was not born at a time of dramatic evolutionary advancements but emerged, evolved, and survived during a period of increased mammalian extinctions. Whereas our ancestors became an effectual part of the African landscape, it is difficult to discern whether their effects were great or small on the rate and pattern of extinction. But there is no doubt that they became increasingly entangled in the web of life.

The perils of interpreting coincidences apparent in the fossil record remain. Close studies of the fossil data, along with computer simula-tions, have closed few options for a plausible explanation of our origins. Quite to the contrary, new possibilities abound beyond the strict cause and effect relations of environment and evolution. Our search for an evolutionary catalyst, a cause of our origins, must venture more deeply into life's chaotic web.

5

Rebels Without a Cause

IT USED TO BE so easy to teach university courses on human evolution. There was a clear succession of just a few known species from *Australopithecus* to *Homo sapiens*—from the time when our ancestors first stood up to the time we grew our auspicious brains. Dart's savanna hypothesis provided students an explanation of our bipedal origins that made sense. Revelations from climatologists that a cooling and drying of Africa coincided with the origins of our upright stance seemed to put a satisfying ecological cap on the explanation.

Then paleoanthropologists found more fossils, and named more species. Paleontologists studied the succession of plants and animals that arose and became extinct throughout the phases of human evolution. And so the picture became more complicated. The subsequent debates arising from the increasing complexity exemplified Washburn's Law: "the less we know about the fossil record, the more confidently we can speak about it."[1] Now students are presented with a barrage of debates and left in a bit of daze. But there must be some way to make sense of the fossil data and understand the machinations of our evolutionary past.

The difficulty started with a small cadre of anthropologists and paleontologists who picked apart the savanna hypothesis bit by bit.

The climatic change, and the evolutionary origins and extinctions that went with it, appeared to be too gradual and continuous to offer an adequate explanation for any sudden appearances of hominid novelties. The hominids were too adaptable to a variety of environments to have their evolutionary course delimited by the slow expansion of one environment and retreat of another. They could fill numerous positions across the web of life and not suffer from the loss of any one particular role or habitat.

By now the cadre of savanna hypothesis doubters has grown to subsume most scholars who study human evolution. But, to be sure, some individuals still hold steadfastly to its neat explanatory power. After all, the expansion of the savanna had to be relevant—it *did* happen, and the hominids *were* there. We cannot ignore that environmental change. But in what ways, and to what extent, might it have been relevant? The *whole* truth concerning the ancestral human relationship to changes in the African environment, among other changes occurring simultaneously, has yet to be found. The rebellion against the savanna hypothesis, and against the turnover-pulse hypothesis as well, has left us without an adequate explanation of the cause or causes of our peculiar evolutionary path. We are truly rebels left without a cause.

Like a phoenix rising from the ashes, a number of ideas are emerging from our new and confounding insights into the past. Some such notions deal solely with the origins of bipedalism more than 4 million years ago. For example, anthropologist Kevin Hunt of Indiana University suggested that bipedalism helped early hominids gather food; he noted that modern chimps stand bipedally to reach small fruits on trees of the open forests and implied that there would have been a selective advantage for our ancestors to do the same, habitually.[2] Peter Wheeler of Liverpool John Moores University took another approach: he noted that an upright stance helped keep hominids cooler by reducing the amount of sun on their backs and by getting them up where a breeze could offer more relief from heat.[3] Both of these notions offer a glimpse into part of the selective advantage of bipedalism and are likely to be relevant considerations. But they are not enough.

Neither Hunt's nor Wheeler's explanation on its own, or any other single cause, seems sufficient to account for bipedalism. There are many other ways to adapt to environmentally imposed dietary regimes and heat stress, as evidenced by the wealth of successful African animals: giraffes can reach high foliage with long necks, elephants have large heat-dissipating ears, and so on. Something more must be involved to account for the peculiarities of the human evolutionary path. As with most explanations, both Hunt's and Wheeler's ideas rely to some degree on environmental change as a catalyst to evolution. Early hominids, they suggest, adapted in a peculiar way to particular challenges brought on by the decline of African forests. This takes us little beyond the savanna hypothesis.

Thus we must go back to basics and find some rock-solid principles upon which we may build a better explanation of our human origins and subsequent evolution. Ideally these principles will account for all phases of the human evolutionary past and fit all the fossil evidence we have to trace our prehistory. These basic principles should also offer mechanisms to explain, or alternatives to replace, the idea that just will not go away—that of ultimate causation through environmental change. We will probably never find any "laws" of evolution; biology, unlike physics, is too riddled with chaos to rely upon any law to be applicable in all instances—that is, unless chaos itself is one of those laws.

In order to find guiding principles, we must first complete our story (fig. 5.1). So far we've seen the origins of bipedalism with *Australopithecus* (and possibly *Ardipithecus*), starting more than 4 million years ago. We continued to trace our evolution through the great hominid divergence, involving the origins of *Homo* and the robust australopithecines by around 2.5 million years ago; this coincided with a gradual lowering of global temperatures and greater climatic variability. Considerable brain expansion started with early *Homo* and continued into *Homo erectus*, increasing the adaptability of the lineage. *Homo erectus* lived side by side with the ill-fated branch of large-faced robust australopithecines until these vegetarian cousins of ours went extinct sometime after 1.4 million years ago. The rest of the narrative, as we can glean from the fossil record, proffers more clues to help us explain our evolution (and perhaps adds more confusion as well). We must build a framework of evolutionary principles that accommodates the full sweep of this story. If the principles hold up to scientific scrutiny over the years, they may once again simplify the task of teaching courses on human evolution.

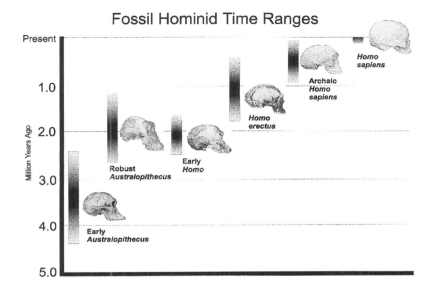

Figure 5.1 • Progression through time of major phases of fossil hominids.

All the World's a Stage

The initial repercussion of early *Homo*'s behavioral and morphological success, after the hominid divergence, was the expansion of our ancestral population. The growth in numbers bred more successes. Early members of our genus used their adaptability to spread out from their aboriginal African home into new environments halfway across the globe—to Asia and, eventually, Europe.

The expansive journey was first undertaken by a few wandering members of the *Homo erectus* "species." They almost certainly encountered new challenges both within and beyond their former African home, and their adaptability allowed them to push into these new frontiers. Their stone-tool technology had improved somewhat, and indeed they were hunting game in order to bolster their omnivorous diet. We have seen evidence from Swartkrans that *Homo erectus* eventually captured the power of fire, forever shifting the balance of nature's power toward human hands. Such are the benefits of having an expanded brain.

Strange, it may seem to us now, how the robust australopithecines with their high-fiber diet went extinct while the hominids with an increasing reliance on animal flesh began to thrive. And thrive they did. The adaptable humans stepped with alacrity into the rich habitats of Asia, seizing opportunities and stretching the bounds of their adaptive capabilities. Certainly this happened by a million years ago, and perhaps somewhat earlier—paleoanthropologists are still grappling with the few fossils we have in order to understand this most fascinating time in our prehistory. More fossils and artifacts, for which we are actively searching, will fill in the few remaining gaps. Nevertheless, some trends of our ancestors' evolution are discernible from the fossils of *Homo erectus,* and these shaped humans as we know ourselves today.

Evolutionary change never stops. Despite the artificial tools that provided our ancestors the ability to manipulate the environment and even to create their own immediate habitats, the forces of natural selection capitalized on biological variants as they emerged. *Homo erectus* was not at all a static species, for the brain continued not only to enlarge but also to become more complex. The functions associated with speech, language, thought, and understanding became lateralized to the left and right of the brain—a unique cerebral division of labor. The face, once adapted for heavy chewing in australopithecine ancestors, gradually became more gracile. Brains had won over brawn.

Such changes indeed made the species extremely adaptable, in both the evolutionary and behavioral senses of the word. By 500,000 years ago, at least in China at the famous site of Zhoukoudian, we have evidence that *Homo erectus* had broken the boundaries of the tropics and subtropics and moved into more temperate regions. Our ancestors' adaptability led them into those seasonably variable environments, but chance combined with natural selection to determine their particular features in each region of the Old World. I have little doubt that some of the biological variability we see today among humans began when *Homo erectus* pressed into these new territories.

There is considerable debate among my colleagues as to what happened next in our prehistoric emergence. The fossil record shows that within the past 400,000 years or so, amid successive glacial advances and retreats, archaic forms of *Homo sapiens* began appearing across the Old World. They varied from place to place, with a mosaic of "primitive" and "modern" traits, but were recognizably among (or at least close to) our own species. Within the last 100,000 years, fully modern *Homo sapiens* filtered across the globe, apparently starting in Africa.

Whether this was a continuous process of gradual evolution or an imperialistic takeover of the planet by one particular group is the focus of the debate.

The matter is far from settled, but I must lay bare my biases for you. As intimated above, I side with those who read a degree of regional continuity from the fossil record of humans. In other words, some of the features found in the skulls of *Homo erectus* within the past half-million years appear to have continued among their human descendants who lived in the same geographical areas; early Africans, Europeans, and Asians who lived a half-million years ago (or less) have similarities to Africans, Europeans, and Asians now. But in all those places there were also parallel trends of brain expansion and facial reduction. I see these parallel developments as a combined result of hybridization between the peoples of various regions along with an overarching natural selection for larger brains. In other words, under this regional continuity model we have long been a unified but diverse species.

An alternative point of view comes from genetic analysis, one of the newer tools in our equipment for interpreting the past. There are those who interpret modern human genetic compositions as relics of the past, much in the same way that paleoanthropologists interpret fossils. By looking at genetic similarities and differences among living peoples across the globe, they establish degrees of genetic similarity and essentially, from the amounts of dissimilarity, clock the times at which different groups of humans diverged as genetically distinct populations. In particular, these researchers study genes not within the cell nucleus, where the genes carry instructions for building a body, but in the cell's powerhouse, known as the mitochondrion. Mitochondrial DNA presumably changes at a regular rate, by chance rather than natural selection, and is passed to offspring only from the mother. To make a long and complicated story very short, some geneticists have suggested that all humans alive today can be traced back to a theoretical woman, a hypothetical "Eve," who lived between 100,000 and 200,000 years ago, probably in Africa. She was not the only female around at the time but perhaps the only one who ultimately left behind her peculiar mitochondrial DNA. If this is correct, then all subsequent humans across the planet must have spread from the original African population, *replacing* all humans who had gone before. Among the peoples who vanished were not only the *Homo erectus* who had settled much earlier in temperate China but also the champions of ice age Europe—the Neandertals.

The methods geneticists use are as ingenious as they are provocative. And the implications of their replacement model, when compared with those of the regional continuity model, are profound. What could have happened to the *Homo erectus* or archaic *Homo sapiens* populations of Europe, the Middle East, and Asia if a single group from Africa was our sole progenitor? What possible superiority could this alleged African group have had over all the rest? Would the expanding African group have exterminated other populations by violent means or simply outcompeted them for resources? Or did we evolve from the African group in friendlier fashion, by the spread of their improved genes from population to population, thus becoming one large but variable species across the globe?

I do not wish to get too far into the debate about the mechanism by which modern *Homo sapiens* originated. Much more evidence is needed from both the paleontological perspective as well as the genetic.[4] The geneticists and the morphologists cannot both be right with their antithetical views, but they can both be wrong. Both have convincing data but fall into the same trap. The geneticists look at genetic variation and try to come up with the most likely scenario that explains the current global pattern: modern humans spread rapidly from a single African source. Their method of choosing this scenario, based on probability analysis, discourages them from looking at the full range of possibilities. Unfortunately, as we shall see, evolution rarely takes the average, predicted course. Frankly, we morphologists may make the same mistake. We see long-term continuity of skeletal traits in particular geographic regions and also opt for the simplest explanation: that of common descent. But, as we know from the complexities of the hominid fossil record, similar traits occasionally show up independently in separate groups. So the simplest explanation may not always be the correct one; newer groups having traits similar to older groups may not actually be their linear descendants.

In my view the geneticists and the morphologists are probably both wrong, largely because the course of evolution and the fate of genes are chaotic and unpredictable. We need a compromise model that can explain both the early establishment of local morphological traits and the late appearance of globally distributed genes. Rather than cling to a single, most probable explanation as seen from one or another biased perspective, we must explore the full realm of possibilities within the bounds allowed by a chaotic evolutionary system. Therein will lie the answer or answers to the way in which *Homo erectus* populations yielded to archaic *Homo sapiens* and archaic forms became modern.

Whatever happened in the past resulted in us. *Homo sapiens* arrived across most of the globe. It was a long journey, from some ape-like creature such as *Ardipithecus ramidus* to my *Australopithecus africanus* friends at Taung and Makapansgat to the nascent early *Homo* and, through the expansion of *Homo erectus*, eventually to us. Along the way, especially near the end, when our ancestors pushed into temperate and downright cold territories, curiosities arose: these included the classic Neandertals, who may or may not belong to our direct lineage. There were undoubtedly many other additional side paths and interesting events along the way that further meticulous analysis of the fossils will bring to light.

The themes at any stage of human evolution, however told, have common features. Each successive stage involved the spread of an adaptable creature into new and varied environments. Better tools and more complex cultures seemed to accompany the transitions, as did the global environmental changes of the ice ages. Adaptability, led by the graces of ever increasing brain capacity, was the hallmark.

We are one of well over a million animal "species" that are "known" to exist today. Humans are quite a successful species both geographically and numerically. We now inhabit most of the planet not covered by ocean and have visited nearly all of it; we have even traveled beyond the planet. When you read this, there will be more than 6 billion people on earth, and many more are on their way. Despite the odds against a few peculiar animals with a bipedal gait who braved the forests and savannas long ago, they survived and initiated a lineage that has now ramified well beyond the original African home. It should be easy to pin down the cause of this success, but it is not. The causation gets mixed up between genetic, morphological, and environmental factors, and these rarely meld together into a consistent whole.

To understand human evolution, we need a theory that is consistent from the environment down to the genes, and from the genes back up again to the environment. We must glean from the evidence what drove the continuous process of human evolution, either because of or in spite of climatic change. And we must comprehend why adaptability was our hallmark, for it would have been so easy for our ancestors to follow a path with quicker payoffs blindly toward specialization—and eventual extinction.

A Royal Blush

While the hominids in the lineage that produced humans were evolving and moving into new environments, other animals were doing the same. And as new species arose others went extinct, just like our robust cousins. In the previous chapter we saw computer simulations in which the origin of any particular mammalian species could occur anytime, by chance, and yet new species appeared overall at a fairly regular rate. (Indeed the only steady, unwavering parameter of the first computer model was the overall rate of species origins and extinctions.) But if evolution actually is occurring by chance, why should it proceed at a regular rate? Then again, even improvisational jazz usually keeps a steady beat; it is intrinsic to the nature of music. Perhaps this apparently constant beat of the evolutionary pulse is a clue that can lead us toward finding a general principle, some sort of driving force intrinsic to biology.

Leigh Van Valen of the University of Chicago is an evolutionary theorist, and a man of many fascinating ideas. He is among the few who use hard data from the fossil record to support innovative theoretical notions. His most widely discussed contribution, known as the Red Queen's hypothesis,[5] strikes at the heart of our problem—the question of regularly paced species origins and extinctions.

Those who are familiar with Lewis Carroll's marvelous chess game, played out by Alice in *Through the Looking Glass*, will have met the Red Queen herself.[6] Alice and the Red Queen ran hand in hand, faster and faster, but the scenery around them never seemed to change. Alice was rather distressed that they did not seem to be getting anywhere. "Now, *here*," said the Red Queen, "it takes all the running you can do, to keep in the same place."[7]

The Red Queen, and Lewis Carroll, probably did not have evolutionary theory in mind, but they inspired Leigh Van Valen's particular theory of evolution. It goes like this: for all living beings, as with Alice and the Red Queen, it takes all the evolving a species can do just to keep pace with nature—to stay in the same "place" while the environment and other species are constantly changing. One would normally think, as did Alice, that running the evolutionary gauntlet should get a species ahead—it should become better and better adapted to its environment, more and more fit in the face of natural selection, with better protection from predators, improved abilities to exploit resources, and so on. But every other species is also jumping evolutionary hurdles just as fast. Relative to one another, then, no species makes any particular progress—predators and prey alike continually move the goalposts. A species seemingly cannot win Darwin's struggle for survival, at least not for long.

Van Valen's hypothesis is simple, and is based on the rather firm notion that species in the wild interact with each other in many ways—in a chaotic, nonlinear system. An evolutionary advance in one species, no matter how large or small, alters the adaptive environment of other species. These other species, who either relied upon or perhaps competed with the first, must then evolve to keep pace with the new regime. Or they might go extinct. The evolutionary alterations of the ones that manage to catch up may then alter conditions for the first species that started this whole mess, as well as for other species who share the ecosystem.

In effect, an arms race ensues. Like the Red Queen, who must constantly run to stay in place, a species must constantly evolve to keep pace with its evolving competitors. In the United States, such a race would be called "keeping up with the Joneses." In nature, it is an antelope quickening its stride to escape the ever faster carnivore, or vice versa. The Red Queen could also be a bacterium that has become resistant to antibiotics, or a cockroach defiantly marching through insecticide. In Van Valen's words, "new adversaries grinningly replace the losers."[8]

Evolution on the Red Queen's treadmill would take place constantly, as long as animals interacted in ecological systems. Much as in the computer models of evolution in Africa, species would continuously evolve and go extinct with no particular pulses or startling events. The engine of evolution would be fueled continuously by changes in any component of the system that, in turn, could spark a change in a related part of nature's machinery. Evolution would be characterized by perpetual motion and a constant rate of change. That would be the source of the steady beat.

One intriguing aspect of the Red Queen's hypothesis is that no outside force would be needed to propel the biological system of evolution. No massive change in climate or geological uplift would be necessary to spur evolution into action. Evolution would happen anyway, for any change in a single being would set off a chain reaction throughout the rest of the food web. What goes around comes around, so even the species that started the series of events would eventually be affected by changes in the surrounding environment.

The Red Queen's system is not totally rigid in its continuity. Occasionally the pace may quicken a bit if a particularly important animal, one that is key to the operation of an ecosystem, evolves some particularly disruptive habits. Changes in climate may also alter conditions and contribute to the system, but no more so than the infinite feedback loop of changes in the biological community itself. Through geologic time and across geographic space, the process would appear regular and constant from the perspective of the fossil record.

From this broad perspective, the Red Queen's hypothesis looks quite appealing. The predictions and the observations fit together neatly, giving the Red Queen and her hypothesis a healthy glow. But when one gets down to details, as happened in the case with the *apparent* neat fit of climatic and evolutionary events, things get a bit more messy and a touch less explicable. The Red Queen stops glowing, begins to perspire, and eventually breaks into an unregal sweat.

Breaking the Imperial Law with Empirical Data

When Van Valen first presented the Red Queen's hypothesis in 1973, it was coupled with what he brazenly called a "New Evolutionary Law." This was the law of extinction, which stated that as life evolves, the environment for any particular species deteriorates at a stochastically regular rate (with random fluctuations). Thus extinction should also occur at a stochastically regular rate. In other words, whereas we cannot count on extinctions happening precisely like clockwork, the overall rate of extinction should vary little when viewed over long periods of time. If only it were true.

In the previous chapter we noted that, starting about 1.8 million years ago, the relative rate of extinction seemed to accelerate. At Swartkrans we saw how a large group of species, particularly large carnivores, seemed suddenly to vanish from the sequence of ancient cave deposits. In East Africa there is good evidence as well that the extinction rate accelerated during the Pleistocene. Some may view this increased rate as a small variation that, when weighed against trends over greater lengths of time, could be expected under Van Valen's "law." Or is it a violation of the law?

One might suspect that the advent of the genus *Homo* and subsequent evolution of our species was at least part of the cause of these extinctions. Certainly this has been true for the past 10,000 years, if not earlier. The success of modern humans and perhaps their predecessors created an unusually severe form of environmental degradation for other species, and they could not keep up in the Red Queen's race to stay in place. Instead of the fairly stable race envisioned by Van Valen, the balance has been upset by humans. Such a sudden shifting from a stable system to one of rapid decline—and there is no doubt that we are witnessing a rapid decline in biodiversity today—is characteristic of nonlinear dynamics. It is chaos writ large, with the chance origin of our own lineage being the small factor, the butterfly effect, that sets off enormously amplified consequences. "Curiouser and curiouser!" cried Alice.[9]

A further criticism of Van Valen's law comes from no less than Elisabeth Vrba.[10] She makes a distinction between the mechanisms of species and population extinctions. Sure, she argues, the advent of a new species in a particular ecosystem can affect other populations of beings around it. But most species occur across many ecosystems, so a degradation of the environment in one part of a species's habitat does not necessarily translate into the demise of the entire species—only of one local population. Or the threatened populations can move, like my great-grandmother, rather than "pay rent." Thus it would take something quite substantial, like the evolution of a widespread, successful species, to lead to extinctions of other whole species. Humans come to mind as an example. But when such a large disruption happens, the rate of extinction increases to the point that it is visible in the fossil record—and the "law" is broken.

The flip side of Van Valen's law, that of species origins, also suffers from a problem of plausible mechanisms. In order to keep the system going, to keep the web of life alive, new species must constantly arise to meet the new challenges. But when the comforts of one's environment are degraded, extinction is more likely than evolution. This argument has been used against the turnover-pulse hypothesis as well.

Most species cannot suddenly adapt to a significant new challenge, be it a change in climate or change in the composition of fellow animals in an ecosystem. Evolutionary mechanisms are too slow, and too dependent on chance, for such quick adaptation to happen. (In the next two chapters we shall investigate exactly why this is true.)

Finally, even if the empirical data supported Van Valen's hypothesis, and even if the required mechanisms for evolutionary change did exist, difficulties remain in applying the hypothesis to explain *human* evolution. There are two reasons why.

First, constant degradation of the environment does not help us explain the timing of particular evolutionary events. For example, if environmental change is continuous, why did bipedalism evolve between 5 and 4 million years ago, not earlier or later? And why did the hominid divergence take off about 2.5 million years ago after a long period of relatively little evolutionary change in our ancestors' lineage?[11] The Red Queen's hypothesis does not help us dissect out the relevant components leading to these events, because it is built on the assumption that such events occur stochastically. This allows Van Valen to focus on the large picture—the overall pattern of everything racing against everything else—without having to explain the underlying individual events.

Second, and more important to the potential worthiness of the Red Queen's hypothesis, the hominid lineage is characterized by highly adaptable animals. Even from the start, hominids did not appear to be at the mercy of any particular environmental condition. They were too widespread (as per Vrba's argument) and too versatile to be affected profoundly by the advent of two or three new species that presented challenges to their survival.

The Red Queen's hypothesis may (or may not) be valid in general, but it has little to say about us in particular; it is not the solution we seek. We are still left as having evolved without a cause. Somehow we must explain the process of evolution in one peculiar group of adaptable animals.

Variability Selection

Rick Potts, director of Human Origins at the Smithsonian Institution's National Museum of Natural History, noted the profound adaptability among hominids and sought from this an explanation of what shapes evolution. He proposed that the grinning victors of the evolutionary race tend to share something important with the hominids: the ability to adapt readily to new environments. Formally his theory is known as "variability selection," but it is best described (in his own words) as "survival of the generalist."[12]

Basically, Potts's idea is that the animals best suited to survive are the adaptable generalists, those not tied down to any particular environmental niche. It has long been a dictum in evolutionary science that specialization tends to lead to extinction, but Potts takes the idea further. He notes that during the late Pliocene and Pleistocene, when hominids were evolving, there was an increase in climatic variations; temperatures not only got cooler, as we noted earlier, but fluctuated more widely. From millennium to millennium, temperatures could vary considerably. Within each year, seasons began to play a larger role. Moreover, it was a time of great volcanism and geologic uplift in Africa. The environment was truly unreliable. Only the most adaptable creatures could survive, and among them were the early members of the genus *Homo.*

These hominids were generalized in many respects. Members of the *Homo* genus retained some "primitive" morphology, adaptively eschewing the gain (or loss) of specialized features for specialized functions, such as the larger jaws and teeth that evolved in the more robust hominids. But was the evolution of a large brain not a specialization? Morphologically speaking, it was; not only was it the focus of much of natural selection, but the brain's development and maintenance harnessed much of the body's resources from birth through adulthood. From an ecological perspective, however, the behavioral flexibility afforded by that large brain turned our ancestors into exceptional environmental generalists, going beyond the environmental generalizations of early *Australopithecus.*

There is considerable merit to Potts's proposal, and it is grounded solidly in Darwinian theory. But then shouldn't survival of the generalist *always* be true? Do we need to impose cataclysmic climatic and geological events on evolving mammals to cause the effect? If one thinks of the web of life continuously changing, as in the Red Queen's world, environmental change is a constant. There would always be animals generalized enough to survive as well as others so specialized that they could be outcompeted by a new competitor or outsmarted by a new predator. Indeed, the hallmark of all primate evolution, back to our earliest ancestors some 50 million years ago, is that they retained generalized features and behaviors. The fossil record is littered with extinct primates that became too specialized. The ones that survived and led to early monkeys, then early apes, and then us were always the ones that diversified their niches in nature.[13] That principle alone, although not a cause of evolution, *allowed* us to evolve—eventually— and must be an important component of our quest for understanding.

The Oxbow Lake Effect

If one were to combine some of the ideas put forth in this chapter, the evolutionary system could be likened to a landscape of rivers and oxbow lakes. Such analogies can only be taken so far, for biological and hydrological processes differ in many important ways. But it cannot hurt to envision briefly the evolution of a lineage as a river flowing across a wide landscape of possibilities.[14]

Up close, a major river seems like a permanent feature of the land. Though the water and sediments in the river are constantly changing, and can rise and fall with changes in the weather, the course never seems to change much. But rivers do change and evolve particular characteristics through time. Those of you who have the opportunity to fly a lot, and who spend your time (as I do) staring out the plane window at the earth below, may have observed some interesting physiographic features of the major rivers that cross the continents. Older rivers, flowing across the wide flat valleys they created, tend to meander a lot—they take on the shape of a slithering snake.

River meanders occur when flowing water takes the path of least resistance. Oddly enough, the resistance is often of its own making, from water dropping its load of sediment along one bank more than the other as the river curves around a bend. The swifter currents swing farther and farther around the impediment, building up an elongating neck of land. But such winding does not last forever. The river may again begin to erode the bank it built, and particularly during floods may cut across the neck of land, carving out a shorter, straighter channel. The loop from the meandering river is then left aside as a crescent-shaped oxbow lake.

In biological evolution, a lineage of successive generalist species is like the wide river, straight and true. Evolutionary specializations are meanders across the landscape, taking the path which for the moment seems to be that of least resistance. Perhaps a specialization involves an animal feeding more and more on a certain fruit that is abundant, eventually losing its ability to consume other foods. When the floods come, in the form of new species origins, the specialized animal becomes the equivalent of an oxbow lake, isolated from a new flow of the river. Like the robust australopithecines, specialized species may persist for some time, but oxbow lakes will surely dry up someday. The extinction of these specialists on the side is the "oxbow lake effect" of evolution.

What causes the metaphorical flood of life that cuts across the evolutionary landscape and strands the oxbow lakes? In Rick Potts's scenario it is climatic change: occasional heavy rains cause the river to rise, pushing the generalists through and leaving the specialists behind. Leigh Van Valen's hypothesis would cast the analogy somewhat differently: the oxbow lakes would be left behind by the constant push of generalist species in the main river current, which are following their new evolutionary courses with a swell of success.[15] Either way, the evolution of a lineage is always guided by the current of adaptive change flowing through it.

But if we go deeper into the analogy, is it not the river itself that causes the meanders? Do the waters rise and shift course primarily because of heavy rains, or because the river is constantly reshaping the terrain, capturing and funneling water toward new channels the river itself has been carving? Perhaps the evolution of a lineage of animals is like the flow of such a river, which determines its own course—catalyzed not by external features but by itself. Perhaps evolution is autocatalytic.

Autocatalytic Evolution

One thing the grand theories of human evolution have in common is that they work from the top down, with external events causing or catalyzing evolutionary events. The turnover-pulse hypothesis and variability selection (survival of the generalist) both pin the impetus for evolution on climatic change. The Red Queen's hypothesis steps a bit closer to the immediate environment of the evolving animal by considering changes in the web of life: the living environment of plants and animals with which a species interacts. Yet in all such cases it is change in a species's external environment, physical or living, that propels evolution. An alternative approach would be to view evolution from the bottom up—as intrinsically driven and self-propelled, or autocatalytic.

The terms *autocatalysis* or *autocatalytic evolution* can have many different meanings to different scholars under varied contexts. Thus it is important to explain how the terms are used here. In physics or chemistry, autocatalysis occurs when change in something is stimulated by one of its own products. Sometimes the Red Queen's race is viewed as an autocatalytic system, and indeed it is—one species changes the environment for other species, and when those other species evolve they further catalyze change that ultimately affects and catalyzes change in the original species, itself a product of the same biotic evolutionary system. But autocatalysis can go deeper, to a level *within* each species. Throughout the next few chapters we shall try to answer the question, Can a species be its own evolutionary catalyst?

The differences in these points of view are illustrated in Figure 5.2. In the top-down approach, the species is at the mercy of strong environmental elements. Climate and geology, as seen by some, bear down from the top and exert the most "weight" on species variability. The plants and animals that make up a species's living environment change with the climate and mediate the effects of climatic change. Their evolution in turn affects the way natural selection acts on the species in question—favoring, say, larger teeth for grinding abundant seeds, or larger brains for the wiles to compete with scavengers—ultimately determining the course of its evolution. Van Valen's model is similar in being driven from the top down, but climate and geology are relegated to a less important role and carry less weight than the community of plants and animals that make up a species's living environment. In either model, or any other top-down model you may wish to envision, something external to the species is responsible for catalyzing its evolutionary change.

The bottom-up model of autocatalysis works quite differently.[16] Change starts within a species, at the level of the genes. If an altered gene makes a change in the morphology, physiology, or behavior of a species, that change has to pass through the rigors of natural selection. Selection starts, however, not with the organism's ability to survive a climatic regime or challenges from other species in its community, but with its ability to survive within its own species. The genetic change must allow reproduction: it must be consistent with

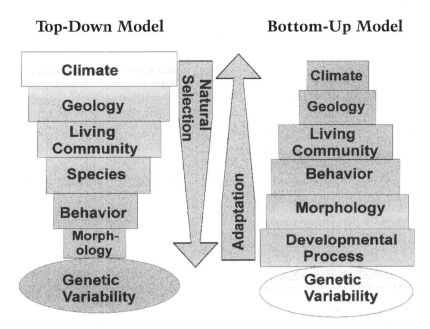

Figure 5.2 ♦ Contrasting interpretive models of the evolutionary process. Natural selection and adaptation work in the same
directions in each model, but the "weight" of various selective influences varies, as represented by the width of the
boxes. In the standard top-down model, outside environmental influences are more important, and the species is
a pivotal unit. In the bottom-up model, genetic variants
must be tested through successive levels within members
of a species before any adaptation to outside influences can
take place.

the morphology of the species, with the birthing and mating processes,
and so on, before natural selection can touch it at any other level.

An individual with the new gene and the changes it brought must
also find food, avoid predators, and survive within the broader living
community. It must be able as well to withstand the challenges posed
by its geological substrate and climatic regime. So these factors do ultimately weigh upon the success or failure of the individual with a particular gene, and thus of the genetic variability within a species. But
step by step, as one gets more removed from the immediate surroundings of the species, each factor weighs less heavily in the process
of natural selection.

The profound implications of autocatalysis can be summarized by a simple thought experiment. Imagine, if you will, the impossible: an animal in a constant, unchanging environment. Under top-down models, without external change, no new evolutionary adaptations would be necessary; a species would be at equilibrium with its environment, and evolutionary stasis would ensue. In the autocatalytic model, evolutionary change would occur whenever chance genetic mutation and natural selection allowed. No external change would be needed (but intrinsic genetic change in and of itself would of course shatter our imagined notion of an unchanging environment for every other living being, and it would send evolutionary ripples throughout the system as in the Red Queen's hypothesis).

Autocatalysis thus resolves one of our dilemmas: we have observed that species regularly evolve while their ecological matrix changes, but it was never clear from this coincidence which is the cause or which the effect. If evolution is autocatalytic, *both* are effects.

What, then, might we be able to say about causation?

HOW DO WE EXPLAIN the evolution of an adaptable hominid? It makes little sense to explain it on the basis of a particular environmental adaptation, such as adaptation to a savanna or a particular climatic regime. Here we investigated theories that deal not with adaptations to specific environmental contingencies but to change itself. We found, however, that evolutionary change is constant and environmental change is constant—so what *causes* the river of evolution to continuously flow and change? An answer may be found if we abandon the search for an external cause and look toward the species as its own evolutionary catalyst.

In the next three chapters we shall explore this fundamentally chaotic mechanism. Starting from the bottom—the genes themselves—we'll work our way up to individuals and species to see how autocatalysis works, and why evolution would never stop even in a static environment. As Alice said in Wonderland, "I know *something* interesting is sure to happen."[17]

6

The Mother of Invention

IT IS OFTEN SAID that necessity is the mother of invention. In a world of creative humans, that is certainly true. With foresight and design, courtesy of our expanded hominid brains, we can artificially shape the world to our needs. Early hominids may have needed something to help them skin an animal, and eventually it dawned on them that a sharp stone, fashioned by bashing one rock against another, would serve their needs. Upon capturing and using fire, our ancestors must have encountered a very sad day when their fire went out. They then needed to recapture or better yet create fire, and eventually invented mechanisms to do so. And in order to keep warm with or without fire, when they needed, they invented clothes. With their stone tools, fire-starting kits, and warm clothes, these resourceful hominids could live almost anywhere.

Today we hominids, *Homo sapiens* to be specific, have long since expanded across the globe, and we interact with one another in increasingly complex ways. Recently it became necessary to keep track of vast amounts of data in commerce, politics, and science. So computers were invented, for such little silicon-based machines are better than our mushy carbon-based brains at storing and quickly processing data. Early hominids needed only their biological computers. We needed something better and could not wait for evolution to oblige, so human ingenuity triumphed once again and met our needs. Since the invention of computers, we have come to rely on them in everyday life, and we use them in novel ways to gain a better understanding of the natural world as well.

In the world of biology, needs are not so easily met. There is no forethought in nature (apart from the remarkable biological instruments within our skulls). And so there is no intentional design in life forms, despite the coincidental appearance of design. Necessity, no matter how urgent, cannot be the mother of evolutionary invention.

Necessity may be the mother of natural selection, in that survival of the fittest "promotes" the traits an animal needs. But natural selection is not a creative force—it cannot invent those traits. It is merely a pruning mechanism, working as well as nature allows with what it is given. The actual force of *creation* in life, in evolution, is much less efficient than purposeful invention and much less directed than natural selection.

That creative force—the mother of invention in life—is chance, not necessity.

Chance, caprice, whim, and fluke have played significant roles in making us what we are. But how can that be? Surely chance alone could not have created something as complex as us? True. There are deep principles involved, but chance still plays a significant role. These principles begin to act at the level of genes, those remarkable storehouses of biological blueprints that are found in every cell of the body. To fathom the biological means of invention we must understand the genes, and how chance affects them.

The Codes for Evolutionary Success

The mechanism by which traits are passed from parents to offspring was a mystery to early evolutionists such as Darwin and Huxley. Today we take genes for granted, and up to this point I have felt free to write about them with very little explanation. Scientists and laypersons alike comment authoritatively on the genetic basis of heredity: we say we have genes for brown eyes, or red hair, or simian folds. Well, not many people talk about the latter. It is a peculiar trait I chanced to inherit from my father, characterized by a single straight fold where I bend the palm of my hand rather than the two curved folds you probably have. Simian folds are usually found in simians (monkeys and apes) rather than humans. But a few of us latter-day hominids are privileged to possess this particular simian trait as well as many other throwbacks, courtesy of our genes. The point here is that the source of many common and uncommon variations of human beings can be found in the genes, many of which have been around for a very long time.

It is worth looking at genes more carefully, to see how they affect our biology and our evolution. Genes, after all, are right at the heart of evolutionary theory, providing the basis for variation upon which natural selection depends. A close look may reveal a few interesting facts about the nature of genes and the means of evolutionary change.

Genes are distinct sections of a long chemical strand called deoxyribonucleic acid, better known as DNA for obvious reasons. Genes organize raw materials for the production of the proteins that build our bodies. They carry the blueprints, in the form of chemical sequences often described as "code," for how to make livers and eyeballs and simian folds. And our genes exist in pairs, with one member of each pair inherited from the mother, the other from the father.

From body cell to body cell, generation to generation, genes cleverly (or so it would seem) pass on their blueprints, using a code that was not cracked by human minds until 1953.[1] To transcribe their codes faithfully into new cells, genes must accurately produce copies of themselves when cells divide. Their very chemical properties encourage just that—accurate reproduction—so that new genes become available for each and every cell in the body (and our bodies are constantly making new cells). Parts of the genetic code must then be turned on or off to produce particular quantities of proteins for gall bladders or irises or skin. It is all quite remarkable, and the genes do their job with truly phenomenal aplomb. Usually.

Sometimes genes make mistakes. Somewhere, anywhere, in a tiny cell of one's body, the code can get mixed up. Too much of this, not enough of that, and the recipe goes bad. If a cell containing the mixed-up gene survives and replicates itself, as cells are prone to do, then the bad recipe becomes a fixed part of the new bodily menu. This process can go haywire and get carried away: cell after cell gets the wrong code, does the wrong thing. Sometimes the wrong thing can be wildly successful and proliferate throughout one's body—it is what we know as cancer.

The body also makes particular cells for passing on genes to new generations. These are the sex cells, the egg and sperm. Like every other cell they too contain long strands of DNA with distinct sections, or genes. But unlike other cells, each egg or each sperm carries only half of the body's coded information: only one gene from each pair of genes. Every gene pair is represented, but only one gene from the pair, inherited from *either* the mother *or* the father, goes into the egg or sperm. The two genes in a pair are often somewhat different: they carry codes for varying traits—a gene from your father coding for brown eyes, say, paired with one from your mother coding for blue eyes. Such alternative forms of particular genes are called alleles.

To understand the nature of genes and alleles, we can look at familiar examples. A gene that controls eye color is fairly straightforward: there are various alleles for blue or brown or green eyes, and so on. So you may have inherited an allele of a gene from one parent that codes for pigment in the iris, and this pigment makes brown eyes. From the other parent an allele for blue eyes may have made it into your genetic structure; people with blue eyes have no pigment, and like the blue of the deep ocean, the blue we see in their eyes is the result of light passing through liquid. If you inherit an allele for blue eyes and an allele for brown eyes, you will have brown eyes. This is because the code for producing pigment is there in every cell of your iris, and those particular pigment proteins will be made even though only half of your gene pair, just one of the two alleles, codes for brown pigment.

Perhaps that was not such a simple example after all, for it gets into the notion of dominant alleles (coding for brown eyes) and recessive alleles (for blue eyes). And frankly I do not want to answer to those of you brown-eyed individuals who have two blue-eyed parents when you cannot figure out where you got that brown-eye allele. It can happen, albeit rarely, given some further complications of genetics (complications that have little to do with infidelity to one's mate).

Genes can work in many ways, some simple and some complex. Often two alleles of a gene in someone's body can average their effects. Your father may be tall and your mother short, but you are somewhere in between—due in part to the blueprints written in the alleles of your genes. Or, one allele may completely dominate while the other exerts no influence, as in the example of eye pigment. Because each gene is inherited somewhat independently from other genes, you embody a chance mix of your mother's and father's traits. Half of your genetic characteristics are derived from your father's genes and half from your mother's. But genes don't code for every detail. Your genetic blueprint is only part of your construction—the foods you eat and the way you live determine every aspect of your biological being, within the range of possibilities allowed by the gene alleles you inherited.

These genes are then replicated in your body's sex cells, and half of them will be passed on to the next generation when and if you reproduce. Your mate contributes the other half. It is truly remarkable how genes faithfully replicate themselves with singular aplomb to be passed on from generation to generation. Usually.

Just as in the other cells of the body, mistakes are made in the sex cells. How often genes bungle the process of copying themselves is not clear. Mistakes are rare—perhaps one error in 100,000 replications of a gene, maybe more, maybe less, depending on the gene. But you have around 100,000 genes that control your development. Chances are good that somewhere along the line a tiny error was made. That chance error in the copying of DNA is what we know as a mutation. Estimates vary, but it appears that each of us may carry as many as four new mutations.[2] The aberrant process of mutation creates new alleles of a gene. New alleles create new living variants—the chance innovations from which natural selection "chooses" the fittest. In the world of biology, mutation is thus the mother of invention.

Whether or not a new mutant allele helps or hurts the survival of an animal, such as our ancestor *Australopithecus africanus*, also depends on chance. In the words of Thomas Hunt Morgan, "In this sense chance means that a variation having appeared, *chanced* to find a suitable environment."[3] Morgan was an early geneticist who, with the help of thousands if not millions of *Drosophila* flies at Columbia University, was instrumental in putting genetic mutation into our concept of evolution. In the early 1900s Morgan understood that the effects of mutant alleles can be good, bad, or indifferent and are usually one of the latter two. For a new mutant allele to work its way into a species, it must chance upon a beneficial effect in two ways. First, the new allele must help in the developmental construction of an organism, be it a fruit fly or a human, fitting in with other genes as a consistent part of the blueprint. Then, in a particular environmental setting, the forms that emerge from the altered blueprint must survive and have some sort of advantage for natural selection to favor them. So they also need some luck.

Oddly enough, many mutant alleles seem to have no effect whatsoever on how well the full ensemble of genes operates. Perhaps a keen molecular biologist could detect a small change in the proteins your mutant gene has constructed, but that small change makes no significant difference to the way your body is built. Before the advent of modern technology to analyze proteins, nobody would have noticed the difference. It certainly would not be "noticed" in the process of natural selection, for the effect of the new allele on your body was neutral. There are many neutral variants of gene alleles passed faithfully from generation to generation, each of which does the job of building a body equally well.

Some alleles produce effects that are far more noticeable but still neutral in terms of survival. My simian fold is quite different from the pattern of folds on your hand (unless you too have chanced upon this peculiar trait). A simian fold may be of mild passing interest to biologists, and a cause of consternation among palm readers, but it in no way affects my ability to survive and reproduce. Its simian origins gave me no specific advantage in climbing trees, for as a child I fell from the lower branches with frightening regularity. Otherwise the fold worked well for my dad and me, and it seems to be of no functional consequence to my nephew, who has it on one hand only. My mother and brothers, like most people, do equally well with their two curved palmar folds.

Whereas many variants among gene alleles produce no selective advantage or disadvantage, sometimes a tiny mistake in a microscopic section of DNA wreaks havoc on the entire organism. At any stage in an animal's development from embryo to adult, a mutant allele may create a situation that is totally at odds with survival. Mutations can kill. Anatomy museums at medical schools often have gruesome exhibits of hopeless monsters created by such genes gone awry. Natural selection is swift and effective in eliminating such genes, for most of the anatomical specimens are of embryos or infants; they never got a chance to live to an age of reproductive potential. Yet the same genes keep popping up in our human population, so new mutant alleles—*the same mutant alleles*—must arise at a fairly regular rate, even if uncommonly.

And yet lethal alleles may not necessarily be weeded out; in some cases natural selection exerts no pressure against them. Killer genes that affect adults only after their reproductive years, sometimes called semilethals, may be passed on with impunity. Huntington's disease, for example, begins to destroy cells in a part of the brain at about age forty, and this degeneration slowly kills the carrier of the Huntington's gene over the following fifteen years. By that time a victim of the disease may have left many children, and about half of them will carry the semilethal gene, which they can pass on to the next generation. True, our knowledge of such conditions today along with genetic counseling may reduce the transmission of semilethal genes, but that bit of artificial selection is fairly recent. In the past, when people lived to a maximum age between twenty and forty, such genes were irrelevant to natural selection. They completely escaped the notice of this otherwise persistent winnowing procedure.

Most mutant alleles are less harmful but still have a bad effect on the poor organism that possesses them. Like typographical errors on a builder's blueprint, these mistakes decrease the efficiency of construction. In other words, the most probable mutation is one that disrupts the biochemical chain of events leading to the development of a piece of our anatomy, or perhaps interferes with the ability of our cells to manufacture a hormone, rather than one that enhances our development and functioning. After millions of years of honing the genes through natural selection, there are more ways for things to go wrong than improve. So the process of mutation tends to follow Murphy's Law: if anything can go wrong, it will. And if the traits affected by these mutant alleles decrease one's ability to survive and/or reproduce, then natural selection weeds out such alleles and discards them in the genetic trash heap. Usually. Natural selection is not always the best of editors, and we are left with many alleles which are less than ideal— good enough for survival but not always fully functional.

But there are a few exceptions to the usual bad or neutral effects of mutations. Sometimes, although not often, biological mutations do something good. They create life in a novel and exciting way. Darwin may not have known the genetic mechanism, but he did not doubt that "during millions of generations individuals of a species will be occasionally born with some slight variation, profitable to some part of their economy."[4] Profitable, yes, but not by design.

Invention, it would seem, is thus the mother of necessity. In Thomas Morgan's sense, the mutation must *chance* to create new life under the right conditions, in which the organism possessing a new trait might possibly find a use for it. If the invention does prove useful, it is favored by natural selection, and with time an animal may come to depend on the new feature afforded by the new allele.

Likewise a typographical error can enhance a sentence and bring it new meaning (albeit rarely). Recently I was amused to note just such a case that came up in a discussion forum on evolution conducted over the Internet. A person who meant to express perceived flaws of evolutionary theory accidentally took the much more enlightened position that "there are wholes in evolution." This led to an interesting discussion about the completeness of evolutionary theory (although there were also quaint comments about "whales" or "whores" in evolution). Whether or not the lexical mutation from *hole* to *whole* was detrimental or beneficial, I suppose, depended upon the readers who constituted the selective environment. But Darwin and Huxley would certainly have been pleased with the "wholes" that have developed in evolutionary theory.

There is no escaping the potential benefits of random mutations, even if the good ones are less frequent than the bad. Mutation is the creative force of evolution. To this largely random process we owe our upright stance, our dexterous hands, and our cerebrating brains. Not bad for little more than a series of chance mistakes.

It may seem that chance could not have produced anything as remarkable as the human body, even if it started with something like an ape. Too much of our construction, despite its flaws, appears to have been designed. But "design" by chance is not as inefficient as it would appear, as long as it is accompanied by a selection procedure. Recently, purposeful designers (of the human type) have taken to using chance mutations as a way to improve the design of anything from efficient assembly lines to jet engines. Their computer-driven procedure employs a "genetic" algorithm. Starting with a particular humanly conceived design, the computer is allowed to "mutate" the design. Most of the altered designs are worthless, making an inefficient assembly line or a jet engine that would not propel even a paper airplane. The computer kills those off (like natural selection, this design process is largely subtractive). But during its testing of new designs the computer selects those few that work well with the mutations chance gave them, and then allows more mutations. Eventually an acceptable design is reached.

The genetic algorithm, involving chance mutation and subsequent selection, works very well for industrial design. Chance can come up with "ideas" that might never have occurred to the limited and convention-bound human mind. Often, in fact, we simply stumble upon our most useful ideas. That is why many discoveries and inventions are attributed to chance meetings or lucky observations. Most fossil discoveries also involve chance, be it a pipe falling out of my mouth onto a fossil deposit or a baboon skull passing along from person to person until it reaches Professor Raymond Dart, leading to the discovery of the Taung child. Chance discoveries can be detrimental to our understanding of nature, as Dart found with the blackened bones he thought were burnt. Scientific persistence, like natural selection, weeds out the good from the bad hypotheses and creates new constructs.

The genetic algorithm also works well in the natural biological world. Usually. A scientific mind then wonders just how it works—how genes can create beings as wondrous as humans and baboons and pigeons and even cockroaches. One also worries why the system sometimes fails—why some apparently thriving plants and animals go extinct or why a particular design is just not as good as one might think it should be. Are there other forces also at work, sabotaging the good effects?

Since the 1920s, when genetics and natural selection combined into the whole of neo-Darwinian evolutionary theory, our vision of the natural genetic algorithm has become ever more clear. We can now see that mutation and natural selection do not stand alone in the continuous creation of life. Science has indeed identified not one but four primary forces of evolution. Knowledge of these forces answers many questions but raises many as well.

The Forces of Evolution

Natural selection, as Darwin taught us, is the main directional force of evolution. It is the source of "progress," in that organisms possessing ever more effective ways of growing and living tend to be selected by the natural means of ever more successful reproduction. Natural selection may get more credit than it deserves for evolutionary progress, but it is the only force of evolution that pushes the parade of life forms toward continued success. None of the other forces are directed toward adaptation, but they are quite important in making us what we are.

Mutation is the creative force of evolution, providing new alleles for trial by natural selection. That creativity, however, has limits. Mutations must work with the raw material provided in the existing genes. So, for example, a pigeon is not likely to hit upon all the mutations at random that would elicit a humanlike brain. Chimps, on the other hand, have most of the genetic prerequisites of a humanlike brain already, and not that many new mutations would be necessary. But aside from such starting-point limitations, this creative force acts at random times in unpredictable ways. We cannot predict when and where a mutation will strike and often do not know what the effect would be if it did. Nature certainly does not know what the outcome will be, so mutations are effectively random with respect to their usefulness for adaptation. Mutation is not a directional force of evolution, although it sets up opportunities for new courses to be pursued through natural selection.

Populations of plants or animals have one more important source of new alleles: other populations of the same species. For example, a population of baboons in East Africa may chance upon new mutations that never appeared among the genes of baboons in southern Africa. But somewhere in between their normal ranges, perhaps during a starry evening on the shores of Lake Malawi, an amorous male from East Africa may encounter a fetching female from the south and pass on the mutant alleles that characterize his own population. Their hybrid offspring may then carry these genes into subsequent generations of southern baboons. This process is known as *gene flow*. As far as the receiving population is concerned, gene flow is a source of genetic variability, a creative force of evolution.

Modern humans are particularly good at keeping up a vigorous gene flow. We have varied populations all over the globe, but with our efficient means of transportation on and above the earth, different peoples are coming into contact with each other all the time. And the contact can often be quite intimate. Thus there is nothing preventing an Australian aboriginal lady from mating with an English aboriginal gentleman, even though their populations have been separated for nearly forty thousand years. A boy from Ohio may mate with a girl from South Africa and produce two offspring to carry on their diverse lineages. It is all part of the happy world of gene flow.

Unlike modern humans, who interbreed fairly freely around planet earth and sometimes above the earth, there are many very small populations of plants or animals that have considerably fewer choices of mates. They may be isolated on a remote island, such as the tortoises of the Galápagos Islands. Or perhaps their continental population just dwindles to a few lucky survivors, as sadly characterizes mountain gorillas and many other endangered animals. Small populations have a problem transmitting a rich variety of alleles to future generations. Because there are so few mates to choose from, and thus not many different alleles of genes around, some alleles become more frequent in the population by chance alone. An adult silverback male gorilla, for instance, may win the competition for mates and establish a harem. Because the total population is small, this one male's alleles become disproportionately common in the next generation. Other alleles may be lost altogether. In either case, whether an allele becomes more or less frequent, its adaptive value plays no role. In a large population, such fluke events have little effect on the overall frequency of gene alleles. If one or a few individuals fail to pass on a certain allele, there are many more potential mates with that same allele. But in a small population, a random event can promote or wipe out a significant fraction of unique alleles. The frequency of a potentially good allele will thus fluctuate randomly to some extent in any population but especially in a small population, and it may even be replaced by an allele of much less adaptive significance. This "selection" by chance, rather than by survival value to the organism, is called *genetic drift.*

In the natural world, then, the random, whimsical forces of evolution have many opportunities to tweak the directed, "progressive" effect of natural selection. Just imagine genetic drift among Galápagos tortoises, inhabiting a series of small equatorial islands off the coast of Ecuador. Charles Darwin noted that the local people of the Galápagos could identify the varieties of tortoise according to the probable island from which they came. Each variety of these giant tortoises had its own distinctive shell markings and other characteristics; they were once divided into as many as thirteen species, mostly from different islands, but now most scientists recognize eleven surviving races.[5]

There are not many tortoises on each island, at least not anymore. Only about five hundred individuals of each variety exist, although the populations of some tortoise races number as many as five thousand. Their numbers were probably greater before we humans arrived on the scene, but these tortoise populations were always fairly small and therefore subject to genetic drift. The effect would manifest itself too slowly to be observed in one human lifetime, of course—and in fact almost everything these animals do tries the patience of human observers. Darwin noted in his journal that they moved slowly but persistently toward food sources or breeding grounds. "One large tortoise, which I watched, I found walked at the rate of sixty yards in ten minutes, that is 360 in the hour, or four miles a day,—allowing also a little time for it to eat on the road."[6] A determined female, after a boisterous mating session, could lay a clutch of about a dozen eggs, of which most would hatch and emerge from under the mud pack where they developed. Then, as Darwin noted, many of the emerging hatchlings would be consumed by rapacious birds, keeping the numbers in check. Only after the luck of surviving another fifteen or so years to reproductive maturity would the larger females breed.

Before I get too carried away by my fascination with the lifestyles of Galápagos tortoises, a description of which could go on for pages, I should complete the detour and get back to the point. Each variety of tortoise is distinctive not because of adaptation to different environments, not because of the persistence of natural selection, but because of genetic drift. Since the populations are small, certain alleles coding for particular carapace patterns *chanced* to find prominence on one island, whereas on another island these large lumbering creatures, in their limited numbers, happened upon a different pattern. Their low odds of survival to reproductive age, avoiding birds that prey upon hatchlings as well as disastrous falls over island precipices, kept their numbers low. But survival value did not affect the patterns that human taxonomists use to divide the populations into races or species. Alleles coding for certain patterns died, for instance, with a plunge over the precipice. Natural selection did not do that. Chance did that.

Back in South Africa I noticed an example of a different kind of genetic drift. From Johannesburg down eastward to the coast of the Indian Ocean, one commonly encounters a fairly handsome but particularly obnoxious bird known as the Indian myna. It is hard to miss their boisterous squawks and aggressive ways. Urban legend has it— and the legend may be true—that the entire population of mynas was derived from a single pair of pet birds brought from India, their native home. Certainly it is true that a very small population of birds was introduced to the southeastern coast of South Africa between 1888 and 1900, became locally established, and within thirty-five years spread along the highways to Johannesburg.

An initial pair of birds could not possibly have carried all the allelic variants of myna genes, for each bird had at most two alleles of each gene. All other alleles would have been left behind in India. Even if a small initial population rather than a single pair came over to Africa from India, that small group would have lost some of the variability existing in the large population of the Indian subcontinent. Whatever alleles these few carried with them would have infused the entire population they propagated. It was my personal suspicion, at first, that South Africans had simply chanced upon immigrant mynas carrying a gene for obnoxiousness, for I could not imagine that the entire subcontinent of India endures the insufferable antics of these pesky birds. According to those who have been to India, I was wrong. Nevertheless, the point is that the South African population does not represent the entirety of variation known from the Indian continent.

The mynas provide an example of genetic drift known as the "founder effect." The South African population may look different, on average, from their Indian ancestors due to the loss of certain variants. As their population spreads and grows here in South Africa, new variants may arise and/or old variants may be selected, and with time the populations on the different continents will look more and more different. This was the principle Elisabeth Vrba suggested as a means of speciation in the turnover-pulse hypothesis: small populations, isolated by the environmental fragmentation wrought by climatic change, initially varied due to the founder effect and then increasingly diverged through mutation and further genetic drift. But can genetic drift proceed at a fast enough pace to cause new species to appear in quick evolutionary pulses?

Certainly genetic drift can be a potent force of evolution in small populations. It can even overshadow natural selection, leading to directional changes that are not necessarily adaptive. For example, the mynas of South Africa may have left behind a beneficial allele, thus being stuck with an inferior morphology or physiology as compared to the parent population in India. Although drift is a directional force of evolution, the direction is not necessarily adaptive. It is simply the change of allele frequencies by chance, for good or ill, and is thus effectively random. Like natural selection, genetic drift is also a *conservative* force of evolution that limits variation—it does not create new variants, new adaptations, or new species but only changes the frequencies of existing variants.

So the directional forces of evolution are *natural selection* and *genetic drift*, one direction determined by the contingencies of reproductive success, the other by little more than chance. The evolutionary forces providing the raw material for selection and drift are *gene flow* and *mutation*, one determined by the contingencies of reproductive success, the other by little more than chance. All these forces had to work together to produce us.

Hypothesis Testing

But how, exactly, did these evolutionary forces produce our ancestors and us? Which ones were most important? And at which points in our evolution? We can find trustworthy answers only by the classic methods of science, observation and experiment. There are many ways for science to experiment with the dynamic forces of evolution. Experimentation with nature, as Galileo showed by dropping objects off the Leaning Tower of Pisa, is a great way to discover new laws as well as to test ideas about how things happen. Evolutionary biologists, however, have a bit of a problem with the short human life span, which does not easily permit the sort of long-term direct observations that would reveal the inherently slow workings of natural evolution.

On the other hand, we can infer some principles of evolution by observing changes all around us—the current events of biological variation—and try to apply our revelations to long-term evolution in nature. Charles Darwin, along with every farmer and rancher on earth, got the scientific wheels rolling. The principles of selection are manifest in every attempt to breed a fatter cow, a more productive stalk of corn, or a faster greyhound. It was indeed this process of artificial selection that formed a large part of Darwin's argument for the mode of natural evolution. Today farmers probably know more about genetics and evolution than they care to admit: breed the best, cull the rest. It is quite simple, to a biological limit. But does it occur in nature? Is there natural selection as there is in selection with purpose? Yes, and from that there is much to be learned.

Natural evolution, slow though it may be, can also be observed directly in nature. Textbooks on evolution are replete with observations on fruit flies, particularly *Drosophila melanogaster,* which breed much more rapidly than humans. Cockroaches would do as well. Most observations of naturally evolving fruit flies are made on remote islands; pity the poor geneticist who has to spend time on Hawaiian islands looking for novel variants of the prolific insect. Yet the specifics of our knowledge still come from the artificial endeavors of less fortunate geneticists who breed flies for a living in urban laboratories that are far less natural. After all, experiments often need very specific controls, unavailable on Pacific islands, in order to qualify and quantify observations of evolution in action.

One can look for principles of evolution in the fossil record as well. Fossils provide data for testing broad notions of evolution, but then we are looking at mere traces of past phenomena. DNA is a fragile molecule, and genes are not the forte of fossil preservation; so the genetic basis of evolution is very difficult to test using fossils. Evolutionary biology, as a good science, must then appeal to all the data sources— fruit fly genetics, agricultural selection, and the fossil record—in order to document and test theories about the mechanisms of evolution. Of course the best means of studying *natural* selection, and natural evolution, would be to sit back and watch for a few hundred thousand years. Instead we catch the snippets we can, but it is like trying to judge an entire movie from a five-second film clip.

Yet there is still a way, with the data we have from various sources, to speed up the process. At the same time we can add experimental controls that would have satisfied the likes of Galileo.

Speeding Up the Pace of Evolution— Once Again, with Feeling

In a previous chapter we tried to speed up evolution at the species level. Hypothetical models including various assumptions were built and then experimented with, courtesy of high-speed computers, to see if the assumptions fit the data. Some assumptions, such as the correlation of apparently rapid speciation "events" with climatic change, were thrown out, whereas others, such as speciation by chance, were supported. Far from proved, but supported. The chance aspect of evolution, however, may garner additional support from more detailed and somewhat more refined experiments. Computer simulations at the level of the genes may reveal some basic principles worth considering. Ultimately nature will have the last word, but mimicking the way new alleles evolve, at a much faster rate than in nature, may show us what to watch for in the slow evolution of life.

Computers breathed new life into evolutionary research, and computer simulations of evolution through natural selection abound. So far Darwinian principles have held up admirably. But this testing of hypotheses is somewhat artificial. By necessity, most simulations make no attempt to mimic nature in detail. They are little more than sophisticated analogies—they illustrate principles but do not conform precisely to reality. In a computer model we want to make things as lifelike as possible but also as simple as possible. These seemingly contradictory goals are resolved by honing from the simulation all details except the ones relevant to the particular questions we want to answer. Like artificial selective breeding, which is also far removed from nature, we can learn something but not everything from such attempts.

Richard Dawkins, in *The Blind Watchmaker*, described what are perhaps the best-known computerized attempts at re-creating natural selection.[7] He started with a simple developmental system that "grew" stick figures of branching trees. His computer program allowed chance mutations, and Dawkins selected results that would morph the trees into insectlike stick figures and even the letters of his own name. The exercise was fun and enlightening, but far removed from nature. His computer model showed that chance, coupled with selection, can produce recognizable figures. Fine, but that is not how nature works. Evolution does not sculpt singular entities on a computer screen, one after another, to be chosen by the program's author. Instead, plants and animals comprise large (and sometimes small) breeding populations with variant alleles coexisting, fluctuating in frequency, flowing in and out—not progressing in linear fashion toward some ultimate goal. We can do better.

Computer models can approach nature much more closely by employing the genetic algorithms that industrial designers have found so useful. Most such attempts deal with simulated populations of computer creatures known as cellular automata. Each automaton is a little lifelike machine in the memory "cells" of a computer, trying to reproduce and fill the available memory with its offspring. So far the procedure resembles real life: animals also have limited resources on patches of earth where their offspring may flourish. When these programs run, the computer essentially becomes infested with cellular automata, like a kitchen full of cockroaches. Just as real animals live and reproduce by a set of rules (so to speak), dictated by genetics and environmental constraints, so do the cellular automata. An automaton, like an animal, might sometimes chance upon a mutation allowing it to deal with these rules in a novel way. Within the artificial world of the computer, all the automata go through a process of digital natural selection; the more effective variants take over more and more memory.

Academic journals are now full of experiments with cellular automata, and these creatures seem to demonstrate a very interesting way of "life." Their evolution within a computer is rife with chaos: the results are often unpredictable and extremely sensitive to initial conditions. Yet, sometimes quite eerily, a sort of order often emerges. Something like a little ecosystem of evolving automata develops within the computer, replete with a seeming balance of nature. But then occasionally, and for no obvious reason, their evolution may be punctuated with an event of major change. Other times things just grind to a halt—not because competitive pressure stopped driving the automata toward further evolution, and not because natural selection stopped working, but because *too many* rules of "natural" selection held the evolutionary process back[8]—a limit suspected by Thomas Huxley. In these experiments, other principles were at work.

Perhaps, lurking within the fictional world of cellular automata, there are basic principles that apply to real life. But it is hard to tell. Cellular automata are fairly remote abstractions of life's reality. If these mathematical games could become more recognizable, their lessons appropriate to real-world biology (if any) might become clearer.

What follows is a start at just that, a computer model originally designed to teach my students the principles of population genetics. The initial results were quite intriguing. As it turned out, there were as many lessons for me as there were for my students. The model took on its own life and exemplified the roles of chance and chaos in evolution, while also helping to clarify the limits to natural selection that Huxley suspected so long ago.

Model Building

Programming life, or at least a simulation of life, onto the cold silicon of a computer chip is not a simple task. It took carbon-based life billions of years to build up a complex system of rules and codes for genes—tiny storehouses of body plans that could successfully undergo the trial and error of mutation, natural selection, and the other forces of evolution. So getting it down exactly right on a computer the first time is a bit much to ask. Simulation also takes trial and error. One must start with some simple basics, such as rules for the inheritance of alleles, but even those can get complicated very quickly, as we saw with the determinants of eye color.

Once the baseline for a simulation is developed, one can ask particular questions and hone the model—make simplifying assumptions (OK, educated guesses)—to be able to answer them. We have three main questions to explore here: (1) How does chance work with natural selection in evolution? (2) Does the evolution of a population's genes involve chaos, as in the experiments with cellular automata? (3) Are there limits to the power of natural selection?

If you will concentrate with me awhile on how to create an evolving population in a computer, we can then use it to test ideas about chance, chaos, and natural selection. But this will not be easy, so be prepared for some mental gymnastics. If you are a bit hesitant to plunge into these details, feel free to skip to the next subsection for a summary.

The basics we start with are people and their genes. In theory the creatures we simulate need not be humans any more than dogs or pigeons or australopithecines, but it helps to envision the biological processes that created us if you think of the simulated beings as humans. How many to start with, however, is another matter. Sometimes people are isolated in small reproductive populations, such as on an island in the Pacific. A model reflecting that situation would be fine if we wanted to experiment primarily with the principles of genetic drift. On the other hand a simulation that could individually track the 6 billion people who live on earth today would push my computer beyond its current limits. So perhaps it would pay to start small, just to get the programming right, and then work up to about 10,000 people. That is about the size of a breeding population of modern baboons, and it is not an unreasonable guess for an early hominid population.

A quick note on what I mean by *breeding population:* it is the number of individuals in a generation who are of reproductive age. Young children don't count, nor do elderly people beyond their reproductive potential. So the breeding population is a subset of the total population, but it is the set that matters most when dealing with genetics. Driving this point home, one of my genetics professors in St. Louis once told our class that "genetically speaking, once you are past a reproductive age you might as well be dead." Oddly enough, the old man died the next year.

Now, those 10,000 or so humans capable of breeding that have been created in the computer need to mate with each other. That is fairly easy, though not as much fun as in real life. The computer can match them for mating at random, or nearly so. It is not quite random, because males must mate with females. So the population must start with half males, half females, and when it is time to mate the natural sort of thing must happen. That is not too difficult to program into the computer's instructions.

Once our males and females have mated, their offspring must be given a gender. It is not unreasonable to assign a 50 percent chance to the child's being male and a 50 percent chance to its being female. Real life is marginally different from that, as slightly more boys than girls are born (especially in my family), but we can simplify matters a bit, as in any computer simulation. The random number generator of the computer can determine the sex at approximately 50-50 and record the result with a digital code for gender.

Now, in any one generation, given the chance nature of gender determination, there may be more of one sex than another. But when it comes time for our digital people to mate, they do so in pairs. A few potential breeders, the excess males or the excess females, are left without a mate. Pity, but such is life in the digital world of mating; they die a peaceful computerized death without passing on any of their electronic genes. The same happens in real life, albeit a bit more poetically. To resolve the problem one could program in a bit of infidelity, but that would confuse all sorts of issues. So the evolving population is strictly monogamous in this simulation. Perhaps that is unrealistic for modern humans, and faithful monogamy is certainly not true for baboons or australopithecines, but there are limits to any computer model. We can still look at principles of natural selection, *assuming* monogamy. After all, nobody even *asked* if the cellular automata were monogamous!

People pass on more than just genes for gender (actually only the males pass on the determining gene for gender). They pass on all sorts of genes, with many possible alleles, coding for eye color, height, and palmar folds. In the computer these are simply coded as numbers that identify the alleles. But each person passes on only one of the two alleles he or she may possess of each gene (either the one inherited from the mother or the one from the father). Transferring allele codes from breeders to their offspring is easy enough to program, but choosing which of the two alleles to pass on from each parent is slightly trickier. Due to the fact that genes sort themselves independently in real life, and thus get passed on virtually at random, the same is done on the computer. The computer's random number generator gets used once again to choose whether one or another allele of a gene is passed on. Although geneticists may grumble about "linked" genes (which means that certain genes in close proximity are usually linked and passed on together), the simulation is fairly realistic yet simple on this point.[9]

So, we create a system of randomly mating monogamous male and female individuals who carry a bunch of genes. How many offspring shall we give them? (Remember, we are only counting offspring who become breeders themselves—the breeding population.) Well, if natural selection is going to work on the new allele combinations carried by the offspring, then the number of offspring who survive to become breeders depends on which alleles the offspring carry. In real life, an allele of a gene that codes for a better brain or a stronger leg may help one survive to the age of reproduction. But it is really reproduction, not morphology, that matters, for that is how the allele gets passed on. Darwinian fitness is little more than reproductive success. So rather than create imaginary alleles in the computer for brains and skin and bones, we can cut straight to what matters: how much the feature specified by a particular allele helps one reproduce. We don't have to imagine the feature. We can simply code for its effect on reproductive success with a simple number, just what the computer wants (and what we want to answer our questions). Thus each allele has a reproductive value representing its contribution to the success in life needed for reproduction.

In order to create a stable population (one that does not grow or diminish in size), every couple should leave two offspring who produce two more, and so on. That is perfect replacement. We can code such a neutral, steady-state tendency straight into an allele by assigning it a reproductive value of 2. Some alleles may decrease the chances of survival and/or the chances of reproduction. A mating couple may only live to have one child; another couple may have three or more. In the computer's simulated world, the value of an allele is directly related to the *chances* of how many children a person is likely to leave. For example, an individual may carry an allele of value 1 and another of value 3 paired together in one of his or her genes. These are averaged to a value of 2, and that is the number of offspring this person is likely to leave—but only if reproductive success were determined by that single gene.

To simplify our understanding, alleles with values of 0 or 1 are considered to be BAD alleles. Why? On average, a couple needs to replace itself with two offspring to perpetuate the alleles carried by both parents. Less than two children is BAD in an evolutionary sense, as the number of their alleles will diminish (a couple having only one child passes on only half the alleles from each parent). Alleles with values of 2 or 3 are considered OK; they represent replacement or slightly better. GOOD alleles are those with still greater values (4 or 5). When the computer starts churning out simulation results, natural selection should theoretically favor OK and GOOD alleles over BAD alleles.

There are many genes that determine one's reproductive fitness. So in the computer simulation, the values of each gene pair's alleles are all averaged. It is as if one can have a good brain and strong legs but be among the 0.8 percent of the population with a congenitally defective heart. Or another individual's heart may be able to pump away forever, whereas the weak legs cannot propel that person away from a hungry roaming leopard, even if he or she is smart enough to avoid the leopard's normal territory. It all averages out in life, as in the computer. Each allele of each gene adds or subtracts from the *probable* number of offspring to be left. In our digitally monogamous world, that probability is averaged even further with the values derived from the mate. So each allele of each gene contributes to the probability of reproductive success, and the individual's overall fitness depends on which alleles appear in the other genes as well. This genetic interplay gives mildly harmful mutations a fighting chance to survive, or to drag down a potentially helpful variant to undeserved oblivion. And that may give us hints about the limits of natural selection.

The averages are not sacrosanct, however. A mating pair, either in life or in a computer simulation, cannot leave 2.576 children. Life is fairly finicky about these things. Offspring come in whole numbers (though my two boys sometimes seem like more). Computers are good at rounding off fractions to the nearest whole, so that is what the simulation does. If a couple's alleles average out to 2.499, then only a pair of children shall they leave. A touch higher score, and three it shall be.

Life, however, has its surprises. Not all our successes or failures can be attributed to our genes. There are many fates that may befall even the strongest and the brightest among us: the untimely appearance of a hungry leopard in one's path, an unfortunate automobile accident, a sniper's bullet, an avalanche of snow on a beautiful sunny day—anything may snuff out just about anybody and send his or her genes to oblivion. Likewise—and I mean to belittle no one in particular—a less fit individual (in the Darwinian sense) who is prone to asthma or cancer may leave four or five children behind. In individual cases, natural selection is not always strictly adherent to the dictates of the genes. So the computer simulation must also allow for natural or unnatural disasters as well as dumb luck. A statistical fudge-factor, operating at random of course, can add or subtract from the probability of reproductive success.

We now have males and females reproducing at random in breeding populations, passing on alleles of genes that add up to a probability of reproductive success, and enduring or succumbing to a few mishaps of life. But where will the alleles of their genes come from?

We must establish existing alleles of genes in our initial population and create new ones as well! For the sake of simplicity we start with two alleles of each gene, each with 50 percent representation in our imaginary population. Their relative frequencies should then change each generation according to whether they are BAD, OK, or GOOD. They replicate faithfully most of the time, but we must also include genetic mutations, the mother of invention. That is how we find out if evolution really works the way we suspect.

As in real life, for each gene there must be a certain probability, a minuscule chance, of a mutation. This chance can be determined from observed human mutation rates, which range from one mutation per gene for every five thousand people to as little as one in a million for a particularly stable gene. For the sake of simplicity I have chosen a simple average of known mutation rates to be used for all the simulated genes.[10] The mutations in this simulation can have any effect, positive or negative, although the latter is more probable—most mutations are BAD, though some are OK and a few are GOOD.

Finally we must decide the number of genes to undergo natural selection in the computerized world. Here we can take cues from nature, then work out a reasonable number of genes needed to answer our questions. One classic example of natural selection in action, among the peppered moths in industrial England, sets the extreme lower limit: in their case, quite strong selection is affecting the alleles of one gene. The black peppered moth blends with the soot on trees and tends to survive, while the white moth is readily consumed by birds that spot it on the bark. A single gene controls the difference in coloring. But goodness knows what is happening to the rest of the genes. These moths still have to survive breathing the polluted air and feeding on the polluted leaves. What if white peppered moths have the gene alleles that deal more effectively with those aspects of life? Selection does not happen one gene at a time.

The other extreme would be to simulate as many genes as there are in the entire human genome. It is estimated that we possess 100,000 different genes. They code for everything from iris color to relative thigh length to development of a bump of gray matter on the cerebrum of the brain. That would be an enormous number of genes (and amount of detail) to keep track of. Quite simply, despite modern technology, a personal computer cannot handle all that information, and we would not know how to interpret the computer's results if it could. We can trim things a bit if we simply want to test basic principles of natural selection.

It is often said that 98 percent to 99 percent of our genes are the same as those of a chimpanzee.[11] Fine, but that leaves a minimum of 1,000 genes to track over a period of 5 million years since our ancestors parted company. I'm sorry, but that is still a bit much for us to concentrate on at once. Let's start small and build our way up to the limits of current computers and our all too human minds. We shall see that 35 genes are enough to be interesting, and 50 become mind-boggling. That many genes can boggle the capabilities of natural selection as well, and bring us closer to an assessment of its effectiveness and limits.

With this plethora of assumptions—complex in the simplest way possible—it is time to set the wheels of our model in motion, or at least the computer disks churning. The simulations are still just a shadow of real biology, albeit somewhat more lifelike than cellular automata. But they sure are revealing about the nature of evolution.

The Population Bomb

If you got bogged down in the details of designing a gene-based simulation, or you unashamedly skipped my bold attempt at creatively portraying the obtuse abstractions of computer programming, then you will be glad to know that it all boils down to this: I created a hypothetical population of people and their genes, assigning fitness values to the alleles they carry for each gene. The fitness values range from 0 to 5 and code for the number of children each couple is likely to have, based on the contribution of that allele to the individual's survival and reproductive success. BAD alleles have values of 0 or 1, OK alleles are coded as likely to produce 2 or 3 offspring, and GOOD alleles weigh in with values of 4 or 5.

In a wave of simultaneous digital intercourses, each male mates with a randomly chosen female, in a strictly monogamous fashion. The number of offspring per couple is determined by the averaged fitness values of all their alleles. Each little cherub then carries half of the mother's gene alleles and half of the father's, including any mutant alleles that might have crept into the population's gene pool, and the process continues through many generations. It's basically just like real life—almost.

The first few runs of the simulation held some surprises, even for the author of the computer program. Just to test it, I started off with a small breeding population of 1,000 individuals. Once the first simulation began, I sat and watched the screen, mesmerized and pensive as it tallied each successive generation being created in silicon. It was with no small sense of bravado that I sired thousands upon thousands of offspring within the computer.

Satisfied that my creation was running smoothly, I went about other business. The computer tallied one generation every few seconds; it was going to take several hours to reach the goal of 10,000 generations, and I had baboon fossils to attend to. But when I returned to see how all my digital children were doing, the process had slowed down considerably. The computer was taking more and more time to get through each generation. Its hard disk, which recorded the characteristics of each individual's genes, was audibly chugging away and indeed beginning to sound a tad stressed.

Something was wrong. The considerable slowing of progress was taxing my computer as well as my patience. A few minutes later the screen flashed what I perceived to be an exasperated error message, and the computer quit. It had run out of space. The simulation had created a population explosion, completely filling the storage capacity of the computer's extensive hard disk. My sense of bravado sank as I realized that all my offspring had met their demise due to a rampant orgy of uncontrolled reproductive success. Perhaps GOOD alleles were not so good if left unchecked.

I should not have been surprised. Darwin's basic tenet of natural selection was that more offspring are created than can possibly survive. Within the first paragraph of his famous 1858 letter to the Linnaean Society, announcing the theory of natural selection, Darwin drew upon the work of Thomas Robert Malthus, an economist who had noted that human populations were expanding rapidly at a geometric rate of increase.[12] A year later Darwin provided the famed example of geometric population increase in the elephant, which he reckoned to be the slowest of breeders. From a single pair of elephants living to the age of 90 and leaving three pairs of offspring, who after 30 years continue the process by leaving their own three pairs of little elephants, there would be 15 million elephants alive after only five centuries, according to Darwin's calculations.[13]

If you test Darwin's mathematics, his numbers may be at odds with your own. But his assumptions were not altogether clear. By fiddling with the numbers, you might convince yourself that Darwin assumed each pair was born every 7 years or so, rather than every 30 years. At any rate, we now know a bit more about elephants and can adjust Darwin's calculations. The longevity of African elephants in the wild seems to be about 60 years rather than Darwin's 90. They become sexually mature at age 10, but the most fruitful reproductive years for the females are between the ages of 25 and 45, with one infant elephant born every 3 to 4 years. So with this information in hand, using a conservative estimate of 25-year-old elephant pairs starting to produce a total of five offspring, one every 4 years, we still end up with more than 9 million elephants after five centuries. Darwin's point may be well taken, even if his simulated numbers were not exactly on the mark.

The point is this: with each pair of breeding individuals leaving more than two offspring, a population will grow and soon grow explosively due to the geometric rate of increase noted by Malthus. Even if each elephant pair in the more biologically accurate scenario left only three offspring each, a single pair would be the honorable ancestors of more than 12,000 surviving elephants after only five centuries. After a little more than fourteen centuries there would be as many elephants as there are humans now (over 6 *billion*), with over 15 billion living elephants by the end of one more century.

Inside my computer each pair of digital humanoids had left either two or three offspring, depending on the selective value of their combined genes, and like a cancerous growth their descendants rapidly sapped all the storage space available, bringing the entire process to an undignified halt. Breeding Darwin's elephants in the computer would have done the same thing, with massive herds eventually overrunning even the largest hard disk.

It became necessary to modify the computer program to prune the population of evolving individuals, no matter how proud I was of their successes. I had to ruthlessly kill off droves of the digital individuals that I had so thoughtfully created, not unlike my approach years earlier to the cockroaches in my kitchen. Such is the nature of life on earth, and so it must be in a computerized simulation of evolution. We do not have 15 billion elephants on earth, nor that many humans (yet), so somewhere along the line a sad, premature death must await a vast number of individuals born.

Getting things right was still a bit difficult, however. With modified restrictions on population growth, time after time the simulated populations went extinct. I was initially as inept at pruning as I was at creating a stable population. It turns out that balancing the growth of a population to fit its available resources (the size of computer memory in this case) is a remarkably delicate task. Nature has the same difficulty and never really achieves a true balance. So I took anther cue from nature and employed a Lotka-Volterra model to impose control and gain at least a dynamic balance.

The Lotka-Volterra model, as used for the simulation, is nature's own version of supply and demand. Let's imagine that the digital individuals in my computer represent, say, *Australopithecus africanus*. They are the supply in this system, at least in the eyes of a saber-toothed cat lurking in the woods of the ancient Makapansgat valley or a leopard prowling the escarpment at Taung. As the population of *Australopithecus* grows, so do the populations of predatory cats, which fatten themselves and their cubs on the flesh of the bipedal critter that cannot run away. Should these cats develop a taste for hominids, however, they might find that the supply of their delight dwindles as the cats' demand increases. Once the cats' food supply dries up, due to a population decline of *Australopithecus*, the predators face two scenarios: they must either find other foods or starve. In either case the pressure is off the australopithecines, at least for a while.

With fewer deaths from predation, however, that natural tendency for a population to increase rapidly, as noted by Malthus and Darwin, allows the early hominid population to rebound. With increased supply comes increased demand, yet another population decline, and so on ad infinitum, as dictated by the ecological observances of Alfred J. Lotka and Vito Volterra.

To implement this automatic constraint on a runaway population, I had to make sure that predator pressure somehow affected every humanoid in the computer simulation, for all mating pairs were living in a similar environment. In my original model some offspring were just killed off randomly due to the monogamous lifestyle programmed into the simulations. This loss of life, or at least of reproductive life, resulted when more males than females were born, or vice versa. During those times of gender imbalance, which occurred by chance, some lonely individuals without mates never reproduced. But if I wanted to impose the Lotka-Volterra system, the pruning had to go further—every mating pair had to lose a certain number of offspring, proportionate to the overabundance of population size, lest the population continue its explosive growth. Even so, some mating pairs would have an advantage over others in the Darwinian race to reproduce their genes. Those with alleles that promoted three or even four offspring would have a better *chance* of leaving behind surviving breeders, and so in the long run natural selection would favor those alleles, and evolution would continue. Meanwhile the harsh pressures on the population would dissipate when the number of individuals had been pruned to the initial starting number, at which time another episode of growth would begin, and so on ad infinitum.

The Success and Failure of Natural Selection

Having sorted out the problem of population dynamics, I was able to create and watch what usually were stable populations. It was time to use the computer program to test a few ideas. Most of what I found merely confirmed what my population genetics teachers had taught me with such devotion. I was pleased that their efforts had not been wasted and that my own students could now see a real-time illustration of evolution in action. But as I experimented with population after population, there were a few extra lessons in store.

Initially, natural selection appeared to work fairly well. That was no great surprise but at least reassuring to one's Darwinian sensibilities. Even when 35 or 50 genes were undergoing selection at the same time, most of the favorable alleles cruised steadily and efficiently to high frequencies within the populations. And the slightest advantage of one allele over another was sufficient to trigger selection. When two different OK alleles (having reproductive values of 2 and 3) existed with equal frequency in the starting population, the allele with value 3 would get to relative frequencies of 95 percent to 100 percent in the population. This usually happened long before the simulation ran through all 10,000 generations. It was as if half of the computerized humanoids were slightly faster than the others and could outrun digital predators, thus living to leave behind more offspring.

I suspected that there were limits to how many genes natural selection could work with at the same time, as indicated by others' experiments with cellular automata. But no matter how small or large the breeding population simulated in the computer (between 500 and 10,000 individuals), all the favored alleles always won out in the end, even if they sometimes had a bumpy ride to high frequency. Thus there seemed to be no limits to the action of natural selection among *existing* variants within a population.

All the favored alleles found their way to evolutionary success, that is, if they had a good start (existing in half the initial population). But new mutations, starting from a frequency of just above zero, had a tougher ride to the top. Some made it, some did not. So the fate of helpful mutations depended on when and in whom they entered the population. With new mutants entering sometime after the first generation, and thus starting from scratch, natural selection had to battle the whims of chance. Only by chancing to win the first few rounds, and thus promoting the new mutation into a respectable fraction of the evolving population, could natural selection then exercise its impressive cumulative powers to ensure that allele's ultimate success.

Recall that most new mutant alleles were BAD but that some had values of 4 or 5—GOOD alleles in this system. Despite the efficiency of natural selection and the high frequency of mutation, only a few of these GOOD alleles established themselves in the final populations. Others were lost despite their high reproductive value. Some new GOOD alleles made it to frequencies in the population as high as 20 percent and yet subsequently declined to naught. We can just imagine how this might occur in real life: an important GOOD allele of a gene, perhaps improving the gait of a bipedal *Australopithecus,* emerges briefly in a population but then declines due to the apparent indifference of natural selection. In fact the allele was simply unlucky. It may have chanced to emerge in a subgroup that was already declining, having hit upon a BAD allele for smaller teeth. Being powerless to select single alleles, natural selection weighed the averages but still found them wanting. It did not favor the more nimble vegetarians with weaker teeth as it struggled to maintain other allele combinations instead, coding perhaps for features such as larger brains.

On the other hand some of the new GOOD alleles caught on for good and made it to frequencies near 100 percent in the evolving populations. It seemed that chance was not only creating new mutant alleles but also dictating which of them made it to prominence and which were condemned to ultimate obscurity and loss. Perhaps there *are* limits to natural selection—limits more perverse than previously expected.

But if chance limits the powers of natural selection, can chaos do the same?

A Population Bombs

What would happen if everything were the same when two identical computer populations of humanoids began evolving—identical except for the presence or absence of one individual? It is not a difficult experiment. Random number generators in computers, which I use to determine gender and mutations and matings, are not totally random. If one uses the same random number seed, the same series of "random" numbers will be produced. So by using the same number seed twice, in two subsequent simulations with identical starting populations, one gets identical results.

But to alter the initial conditions ever so slightly, I forced one little change in the starting population. I killed off one individual. It happened to be a female, and I chose her so that her loss would not affect very many of the matings of her generation. Just one innocent female out of a population of 10,000, snuffed out for the cold experiments of science. (This caused some consternation in a colleague of mine who, while sitting at the computer in my lab, looked down to see a note I had written: "Killed off one female." Fortunately it was not too difficult to allay his worst fears.)

I expected that, with the loss of this female, the final allele frequencies of the 35 genes undergoing selection might turn out a little differently. And perhaps a GOOD mutant allele that had not succeeded in the first simulation would now make it to prominence, or another one that had made it in the initial simulation would be lost to obscurity, having been overshadowed by an OK allele. In other words I expected a touch of chaos, a soupçon of sensitivity to initial conditions. But I was wrong.

After starting the simulation one evening, I came back the next morning expecting my computer to be chugging away, recording generation after generation. But it had stopped. Cursing what I thought must have been an overnight power failure in the building, I prepared to start again. But then I noticed that the simulation had come to a natural halt after just 321 generations. What had happened?

By killing off just one individual out of 10,000 in a simulation that led to highly successful evolution through natural selection, I found no *minor* difference. The loss of one individual—*just one*—led to the *complete extinction* of a population 321 generations later. Could life be so fickle? Was this yet another example of a chaotic blue sky catastrophe?

Perhaps I had made a programming error that resulted in this startling conclusion. Just to test whether my program was causing a problem, I tried killing off a different individual. This time, as when I first ran this simulation, my computer chugged through to completion, and the population survived for 10,000 generations. So the simulation technique was not in error but instead contained lessons. Profound lessons, I was beginning to think.

My mind went back to the incredible and disconcerting extinction of a whole population due to the loss of a single, undistinguished individual. One female out of a population of 10,000 could not contribute her genes to posterity, and thus was lost an entire future.

The implications always bring a wry smile to my face. I look back to my seven years at Taung searching for the remains of a relative of the Taung child. But just think: had one more individual been lost, had one more *Australopithecus* prematurely committed its bones to the tufa caves of Taung, the entire hominid population might eventually have been lost. One more death at Taung, nearly three 3 million years ago, and our species might not have been here to ponder its past.

These computer simulations record only one sliver of time. Just imagine the chance events, the genetic faux pas, that could occur over the 5 million years of human evolution, the 50 million years of primate evolution, or 4 billion years of life on earth. Yes, we are lucky to be here, as is every other living being, lucky to have survived the chaos of evolution. And we would not be here without it!

Sensitivity to initial conditions is the signature of chaos. Certainly the loss of a single individual out of 10,000 qualifies for a minor change in initial conditions. And given that the entire population was lost prematurely, she must have been an incredibly effective instigator of the butterfly effect. But she was not a particularly notable digital person, for my computer recorded the mundane characteristics of her genes—not an unusual assortment of GOOD, OK, and BAD alleles. One might think that the loss of her GOOD genes led to the demise of her survivors' descendants, who were not so blessed. But that was not the case. Quite to the contrary, she was nothing special at all when she made her initial contribution to the successful outcome of the first simulation. Yet her loss was just enough to tip the scales of the delicate balancing act performed in the simulation of life as we know it.

One more counterintuitive result emerged. Upon inspecting population trends evident in the reams of data created by the simulations, I found that the loss of our one female—call her Madam Butterfly if you will—led to a slightly *greater* initial success of the population. Madam Butterfly may have been lost, but her survivors went on to produce more progeny than if she had lived. There is just one problem with that: a more successful population elicits stricter controls on further population growth due to the Lotka-Volterra rules programmed in to stabilize population size. It was as if a population of saber-toothed cats multiplied in response to the plenitude of easy if not tasty catches among large numbers of feeble *Australopithecus* individuals. Eventually there were too many cats ready to gobble up the last *Australopithecus*—and they did.

Given the number of species that have gone extinct in the past, terminating millions of years of a lineage and leaving no evolutionary descendants, one wonders if perhaps they were too successful at some stage of their existence. And aren't we humans rather successful now? One just wonders. . . .

Que Será Será

In the next run of the computer simulation, starting again with same population *including* Madam Butterfly, I killed off a different individual. (I assure you that the choice was purely a matter of programming pragmatics, but when my colleague saw a note next to my computer saying "Killed off another female," he again wondered if I was up to no good.) Her effect was not quite so profound, but she had a butterfly effect nonetheless. Some of the really GOOD mutations that had made it to prominence in the first simulation ended up with frequencies near zero. Others, which due to the exigencies of chance had never quite taken hold in the initial simulation, were now quite prevalent in the evolved population.

Thus the exact path of evolution depends on the precise nature of the initial conditions, down to the existence of a single individual. As Darwin foresaw, "If any single link in this chain had never existed, man would not have been exactly what he is now."[14] Essentially the chaotic system, due to an alteration of one initial condition, happened upon a different attractor for the genetic pattern of the population.

It was as if, for example, one population had struck upon genes that coded for slightly larger teeth. Large grinding teeth are beneficial to any vegetarian, such as early *Australopithecus*, for they do not wear down so quickly and allow for the processing of a lot of food. (Later on those teeth might inhibit rather than help adaptation—but that would be another story, played out beyond the 10,000th generation.) Another population may not have chanced to establish the large-tooth genes among themselves. But other mutations were possible, and having chanced upon the untimely loss of one innocent female, this other population proceeded on a slightly different pathway. The direction of that path eventually became more than slightly different and led to opportunities for new alleles and allele combinations, otherwise unforeseen in the former evolutionary path, to establish themselves in this other population. In this case the mutant genes prolonged and enhanced the development not of teeth but of the cranium housing the brain. The brain then became slightly expanded, offering not the ability to grind more vegetation but the possibility of thinking about novel food sources.

The two computerized populations, one developing large teeth, the other large brains, had identical potentials. They initially had the same genes, save those of one individual, to embark upon their respective evolutionary courses. This closeness of starting conditions was a key part of the computer program. And after the starting generation mated, the two populations evolved under identical conditions of natural selection, again as programmed according to rules and principles, with Darwinian fitness through reproductive success dictating the evolutionary course of simple allele frequencies. Yet despite the same principles and rules applying to nearly identical populations (9,999 out of 10,000 individuals being precisely the same), remarkably different courses of evolution ensued. Sometimes things happen for seemingly illogical reasons, despite the imposition of strict principles. Chance and chaos in a *unitary* environment, not varying adaptation to a changing or fragmentary environment, made the populations different.

So the converse of the early hominid prospects after losing the Taung child may have been true as well. If we were to set back the clock and allow that single *Australopithecus* child to live and contribute his genes, rather than die young and be studied by Professor Dart, our human population may have evolved quite differently—or not at all.

But all these experiments so far have occurred only within an imaginary computer world. Intriguing experiments with bacterial populations, each allowed to evolve independently under identical laboratory conditions, suggest the same principles of chance and chaos may apply to living organisms, for each population evolved in a separate way.[15] So perhaps the same principle really did apply to evolution outside a computer. Perhaps natural populations, such as *Australopithecus robustus* and *Homo habilis*, diverged because of chaos. Perhaps robust teeth or expansive brains were the products of chancing upon the right genes at the right time, rather than being the keenly honed products of natural selection in different environments. Perhaps natural selection, albeit efficient and powerful once it gets enough variants to work with, remains fundamentally at the mercy of chaos.

Perhaps too many of these written musings have been prefaced by the cautionary word *perhaps*. Such is the nature of exploratory science.

Drifting into Uncharted Territory

Each new run of the simulation brought new insights. One such insight helps to clarify the limits of natural selection. Certainly the computer revealed no limits of natural selection choosing among *established* genetic variants. In a population of 10,000 or a population of only 1,000, any favored allele was nurtured to prominence as long as it had a healthy representation in the initial population. In the real biological world, this means that natural selection can work very efficiently, for example, during times of environmental change. If *Australopithecus* populations had to adapt their teeth, their limbs, and their nose shapes to survive in drier climates with limited food resources, it would not have taken long. In the computer runs, most selection of existing variants takes place within the first 1,000 generations of the 10,000 simulated. But this quickness of genetic adaptation assumes that half of the population already has the currently favored features. All we would see in the fossil record would be a final loss of primitive features in the teeth, limbs, and nose, but nothing startlingly new.

Again, Darwin predicted this without the benefit of a computer simulation. In fact we see it happen in the world around us all the time, such as when the peppered moth populations of Europe adapted their camouflage to the black soot of the industrial revolution. Natural selection works well *with existing variants.*

As noted earlier, Darwin envisioned more than the selection of existing variants. He also envisioned some sort of ongoing mechanism for *new* mutations among species, stating that variants would arise that were "profitable to some part of their economy." He continued: "Such individuals will have a better chance of surviving, and of propagating their new and slightly different structure."[16] The word *chance*, however, is the operative word. It is this chance (and hence a bit of chaos not foreseen by Darwin) that not only supplies natural selection with variants but also limits the effectiveness of natural selection. The computer model demonstrates this nicely. Even in large populations strange things happen, and natural selection does not provide the sole explanation.

New mutations, no matter how beneficial, only stand a *chance* of finding their way into a population. It is clear that not all beneficial mutations make it to prominence. Whether or not they make it—having first survived the test of fitting in with bodily development—depends on at least two factors: the population size and the number of genes undergoing selection. The effect of population size will surprise no experienced geneticist. Over the same number of generations, a computer population of 1,000 individuals accumulates fewer beneficial or GOOD mutations than a population of 10,000 accumulates. This may come as a surprise, however, to those who believe that new species tend to arise from small populations. It is true that small populations may differ from the parent population and from other small populations due to genetic drift (largely by allele subtraction because of the founder effect), but they must eventually struggle to adapt with new features. Among fewer individuals, fewer new GOOD alleles will appear; indeed, any novel mutants that creep in through genetic drift are more likely to lead to extinction than to adaptation[17]—BAD mutations, after all, are more common. Evolutionary novelty that is useful rather than harmful arises more easily in *large* populations, and success breeds more success. The rivers that capture the most water are those that continue to flow.

The effect of the number of *genes* undergoing selection, however, is a factor that geneticists tend to overlook. As I raised the number of simulated genes per individual that were being selected, I noticed a decline in the number of new mutant alleles that became well established in evolving populations. After 10,000 generations, even with the vast opportunity available in a population of 10,000 individuals, fewer than half of the available GOOD mutant alleles became established when 50 genes were being selected simultaneously. With only 10 genes undergoing selection, not a single GOOD allele was missed. Paradoxically, a larger number of items on the menu for selection becomes restrictive, not creative. Thus, while *Australopithecus* was evolving, all sorts of potentially good adaptations would have been missed if even only 50 of that hominid's roughly 100,000 genes were targeted by natural selection at any one time. As with the cellular automata, and as Huxley predicted, there were surprising limits in these simulations to the ability of natural selection to establish new variants.

Stuart Kauffman noted this phenomenon in his computer experiments and referred to the logjam as a "complexity crisis."[18] Sometimes too many selection criteria thwart the evolutionary process rather than propel it. The complexity crisis in my own simulations showed a further nuance: with yet more genes undergoing selection, the results became increasingly unpredictable.

Even in large simulated populations, one could see the rises and falls of new mutant alleles without any apparent direction. Indeed, if there were a lot of genes being selected, some of the mutant alleles exhibited the characteristics of genetic drift—random fluctuations due to *chance*. As Kauffman noted, sometimes evolution happens despite natural selection.

And—and this is a very important and—the simulations yielded different genetic patterns every time. Natural selection was largely random selection. Sometimes it would be one GOOD mutant allele making the grade, sometimes another, even if the starting population had the same allele frequencies and the potential mutant alleles were the same. That alone is interesting, for it means that changes in a population are not necessarily the consequence of adaptation alone. Given another chance to run the same evolutionary course, a species may develop very different new features. It all depends on the peculiarities of initial conditions. It all depends on *chaos*.

The results make sense if you look at the predicament of animal breeders. All in all they have done quite well in the past 10,000 years or so, breeding corn that gives us many nutritious kernels, cows that produce an abundance of milk, and pigs that harbor a store of tender and tasty meat. The breeders focused their selective powers on a few desirable features derived from many animals. But eventually the limits of selection were imposed. Corn can yield only so many grains, and only if the fields are properly fertilized. Cows produce a lot of milk, but they have limited resistance to disease and frankly are not nearly as canny as their undomesticated bovine relatives. Pigs may present us with delicious pork and ham and bacon, but few would survive if tossed back into the wild. One cannot select for all things at once—there are limits to natural and artificial selection.

Likewise, even after 200 million years of evolution, during which the four-chambered heart became standard issue for more and more species, one in every 125 living humans has a heart defect (and one in every 37 stillbirths has a congenital malformation of the heart). One would have thought that natural selection could eliminate all the BAD heart genes and furthermore could have evolved GOOD alleles to protect the heart from various congenital defects of nongenetic origin. That is not what happened.

Darwin, to give him his due, was not unaware of these limits. He conducted numerous breeding experiments with plants and was constantly frustrated in his attempts to get what he wanted. In exasperation he stated, "all nature is perverse and will not do as I wish it."[19] Darwin was no more efficient at selection than nature, for chance and chaos play prominent roles in the theater of evolution.

GENES CARRY the basic instructions for the building blocks of the human body. Alleles are variants of genes, ultimately determining the variations on which natural selection acts. New alleles arise by mutation—by chance errors that create novel building blocks. Mutation is the mother of invention, and invention is the mother of biological necessity. Larger populations carry more mutations and are more likely than smaller populations to propel useful variants through natural selection. Thus an autocatalytic feedback loop may ensue—a gene catalyzes an adaptation for a successful population, the population grows, and its greater size increases the chance of finding new mutant alleles for the next round of novel adaptations. On the other hand the system can go awry, as we found with the chaotic influence of Madam Butterfly—too much success too soon, and the forces of nature can turn on the evolutionary process.

Whereas natural selection is a necessary force of evolution, it is not a sufficient one. Chance and chaos provide the fundamental driving force and also help determine the direction of evolution—due partly to their role in mutation, partly to their imposition on the power of natural selection to establish new alleles. There are limits to natural selection, and limits to the evolutionary process. A series of computer simulations demonstrates this, but life itself is the proof. As we continue to build the human species from the bottom up, and move from genes to morphology, our own bodies may reveal the nature of our evolutionary origins.

7

"You Can't Always Get What You Want..."

...but if you try sometime, you just might find you get what you need." One can be reasonably certain that when Mick Jagger and Keith Richards wrote the words for their classic Rolling Stones song sometime in the late 1960s, they did not have evolutionary biology in mind. But few phrases could better portray Darwinian evolution, conceived well over a hundred years earlier. Evolving animals often get the adaptations they need but rarely find what they might "want." This cornerstone of biological thought is often overlooked or forgotten.

Just think what weird and wonderful forms life could take if nature allowed species to evolve precisely the traits that they might want. For example, a baboon might benefit from an additional eye in the back of its head to see those prowling cats that have been catching and dragging baboons into caves for the past 3 million years. For other species, convenient dietary habits might evolve: an acquaintance of mine often talks about the wonderful feasts that would be available to a grass-eating snake. On the other hand a blade of grass that grew into a noose and could tighten around the snake as it tried to slither away might be an equally convenient mechanism for grass to obtain fertilizer.

I have often thought that the ideal animal would be a green bird that could photosynthesize, for it would just have to soar beneath the sun to get its life-giving energy. If such a bird were a sentient being, it could enjoy the wonders of the earth and the air, and with such a vast source of sustenance could live in harmony with other creatures as long as it were smart enough not to overpopulate the skies. An evolutionary imagination can run free.

Evolution, which works by chance rather than design, has not done badly, however. Although there are no intelligent photosynthesizing birds—yet—there is a brilliant array of wild and glorious adaptations of which nobody could have dreamed. Green insects that look precisely like leaves gave me the idea for a photosynthesizing animal; something has held them back from that remarkable step, yet their adaptations are wondrous enough. I marvel at the complex webs spun by spiders to catch their prey, the echolocation abilities of bats, and thousands of other unique and seemingly ingenious adaptations. Through its genetic algorithm, nature is infinitely more creative than you or I.

Yet plants and animals often get only what they need to survive, rather than what they might want to thrive. Evolution proceeds in a chaotic manner without a particular strategy or goal. Thus there are incomplete adaptations such as the "flying" squirrel, which cannot really fly, or even maladaptations taking the moth to the flame. Geneticist Theodosius Dobzhansky is well known for stating that "nothing in biology makes sense except in the light of evolution."[1] The truth is that some things don't make much sense anyway.

Human beings provide a perfect example of inadequate adaptation. People often assume the contrary, and there are many books written to extol the virtues of the human body. This is not one of them. Much more can be learned if, for a change, we concentrate on our biologically inherited faults. Such an exercise, depressing though it may be, will help us understand two important principles. First, we shall see the limits of natural selection that arise due to morphological constraints—the ever changing sizes and shapes of parts that form a chaotic web called the "body." (Indeed, the evolutionary interactions of the body, behavior, and culture among humans form a nonlinear dynamic system.) Second, we should begin to see how the connectivity of the different body parts plays an important role in evolutionary autocatalysis.

Four Legs Good, Two Legs Bad

Our upright posture may have carried our large intelligent brains to our self-professed pinnacle of evolution, but those brains rarely think about what went wrong in the process. The late Wilton M. Krogman, a leading forensic anthropologist, thought about it a lot; back in 1951 he dwelled at length on the negative consequences of our idiosyncratic form in an article titled "The Scars of Human Evolution." In his opening remarks he very effectively introduced his approach: "As a piece of machinery we humans are such a hodgepodge and makeshift that the real wonder resides in the fact that we get along as well as we do. Part for part our bodies, particularly our skeletons, show many scars of Nature's operations as she tried to perfect us."[2]

One of the hallmarks of human evolution, as Raymond Dart discovered with the Taung child, is our upright posture and bipedal locomotion. Unlike other primates and most other mammals, we habitually walk on two legs. Some millions of years ago our australopithecine ancestors stood up and set us on a course from which there was no return—despite the problematic consequences. There is a cost or a compromise for every evolutionary advance, and bipedalism is no exception (see fig. 7.1).

Let's start, as did Krogman, with a look at the vertebrae, the building blocks of our "backbone," or vertebral column. This vertical supporting structure does work, but not nearly as well as the horizontal structure of a quadruped. Krogman cited William King Gregory (the great primatologist and first American scientist of note to accept *Australopithecus* as a hominid) in characterizing the quadrupedal vertebral arrangement as a "bridge that walks." Four-legged animals possess a beautiful arch system, the product of tens of millions of years of evolution. A few years ago I read about an award-winning architect who designed a clever highway bridge based on the vertebral structure of a dinosaur. But the human vertebral column is a poorly improvised adaptation, an architectural flop worthy of no prize. In fact we pay a significant price in terms of strained muscles, slipped disks, pinched nerves—the list goes on.

Our common back problems are directly attributable to the evolutionary consequences of taking the superb mammalian arch structure and warping it into a crude pylon of bony pieces. The result was an S-shaped contraption leaving us head and shoulders above our quadrupedal cousins but with vulnerable support. To achieve upright

Anatomy of a Biped

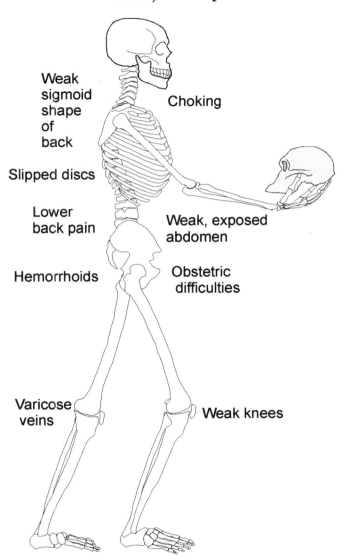

Weak
sigmoid
shape
of
back

Choking

Slipped discs

Lower
back pain

Weak, exposed
abdomen

Hemorrhoids

Obstetric
difficulties

Varicose
veins

Weak knees

Figure 7.1 • The human frame has often been viewed as a crowning evolutionary achievement. There are, however, a few problems
with our upright stance.

posture with this wavy backbone, the vertebrae had to become wedge-shaped; hence they easily slip on each other. Alternatively the disks between successive vertebrae are prone to squirting out as the weight of the upper body presses down upon them. What kind of "adaptation" have we inherited?

As a paleontologist I deal mainly with bones, but it is in the soft tissue inside us, the viscera, where the consequences of our upright stature and locomotor habits really wreak havoc. Our gut, for example, hangs in a most peculiar fashion. Despite the various adjustments that evolution afforded us after *Australopithecus* took those first precarious steps toward a commitment to bipedalism, the attachments of our intestines did not change much. In a quadruped the gastrointestinal tract is suspended from the vertebral column by a double-layered sheet, the mesentery, like curtains hanging neatly from a rod. Our curtain rod has had its orientation abruptly altered for our vertical stance, and now our guts flop down all over each other in a terribly untidy manner.

It is not the untidiness that worries me, and millions of other humans, but the flopping and upending of the gut that results in increased pressure down at the rectal end. I am not alone in suffering from the consequent hemorrhoids. One wonders if *Australopithecus*, without the benefit of 3 million additional years of potentially corrective evolution, also sat and pondered the origin of this irritating pain in the rear end. Personally, I wish that evolution had resolved this problem before my unlikely birth, but you can't always get what you want.

Standing up does not solve my problems, for gravity still pulls things down. My occasional beer belly pushes out on my abdominal wall and edges over my belt (just slightly, I assure you). That abdominal wall has three layers of muscles and some very tough supportive layers of fibrous tissues, but a weak, triangular gap remains near the base through which one's gut can inconveniently and quite painfully protrude or herniate. Like many of my bipedal friends, I have had to endure surgical repair of such a hernia; I now have a real scar to commemorate the evolutionary scar of bipedalism.

In modern society we do not tend to think about this much, but our ancestors' vertical abdomen must have been terribly vulnerable in a wild world full of carnivorous predators. At nature reserves one can often see a lion devour the relatively protected abdominal contents of an antelope, the male usually preferentially starting with the liver. Even the normally docile baboons I watch at Taung and Makapansgat occasionally attack and devour the passing goat, again starting with the rich contents of the abdomen. How exposed and tempting *Australopithecus* guts must have been to a hungry carnivore. One must wonder how our ancestors survived against such odds, with such peculiarly protruding anatomical bits. And frankly, if I had to walk across the savanna, I would not want my genitals to be dangling unprotected during a trek through the sharp, dry blades of the tall savanna grass.

When we bipeds stand, with our abdomens blatantly exposed and chest struggling to stick out above, our hearts are usually about 1.3 meters off the ground. Thus most of our blood flow is vertical—up and down, down and up. To circulate through our highest and lowest extremities and flow more than one meter back to the heart, our blood has to overcome a lot of gravity. Other animals, down on all four limbs, have a largely horizontal flow of blood, with no problem. But we humans need a series of makeshift valves in our veins that help force blood back up to our hearts. It is quite a remarkable adaptation, when it works. But because of our erect posture, those valves frequently break down under the immense gravitational pressure, and we get varicose veins in our legs.

Do you think that other animals get varicose veins, including the particularly uncomfortable and malpositioned form known as hemorrhoids? No, they are better adapted for circulation than us upright, bipedal animals. We took their sensible system, upended it for our striding locomotor habits, and now pay the price.

Evolutionary fitness is defined in terms of reproductive capability. Obviously the more offspring one has, and in particular the more viable and fertile offspring, the more genes one passes on to the next generation. So it would seem that humans, as the self-perceived pinnacle of evolutionary success, would have biologically mastered reproduction. Any pregnant woman would probably assure you otherwise. The added pressure on her upright circulatory system, which commonly results in varicose veins, is among the least of her worries.

There is probably nothing more natural and more exciting in human life than the birth of a baby. But only in humans is this beautiful process so difficult. Other animals can have complications, it is true, but most females of other mammalian species are not nearly as immobilized and traumatized as their unfortunate human relatives. Our difficulties in childbirth are largely a product of the evolutionary expansion of the brain in a bipedal organism. The vastly enlarged human cerebrum has given us many practical disadvantages as well as evolutionary advantages. In short, we have to get a very large head through a very small hole.

If a large brain is good—and it seems to have been a substantial success in many ways so far—then it must be accommodated at birth. One solution could be to enlarge and widen the female pelvis, but that would defeat the millions of years of evolution it took to refine the bipedal gait. A wider pelvis would inhibit the unique striding capability we have achieved, replacing it with a shifting, off-center balance that characterizes other animals during their occasional bouts of bipedal adventurism. I have seen baboons, for example, pilfering bits from our garbage pit at Taung, clutching their spoils with the upper limbs while swaying back and forth on their two imbalanced legs. This precarious locomotion clearly is not a solution for our long-suffering human females.

In human evolution, part of the problem was solved by limiting the gestation period to about nine months: human babies pop out before their brains grow too large to squeeze through. In addition the infant skull was equipped with flexible, soft cartilaginous portions that could temporarily conform to the peculiarly small and confined opening of the birth canal, hence the peculiar conelike shape of the newborn head that many first-time mothers find rather disquieting.

The natural ability of women to stretch muscle and skin to dilate the vaginal opening (a painful consideration of birth often forgotten by those of us who focus on fossil skeletons) is a remarkable testament to the placated evolutionary needs of our species. But it is hardly what an average woman might want. The principle evoked from the words in the title of this chapter is vindicated every time a child is born.

Now, the nine-month gestation period happens to be short, relative to the amount of physical development needed to fashion a potentially bipedal infant with a high cranial capacity. The strategy of giving birth before the fetus grows too large leaves us with the burden of altricial young—those with a long period of dependency. That may be a joy today, for the helpless infant becomes the focus of all the love and attention the human spirit can muster. My two boys certainly bring me great pleasure (along with a sore back). But can you imagine in the past what our ancestors had to put up with as they foraged in the wild? First the women had to waddle and lumber around with these cumbersome parasitic fetuses in their bellies, while trying to find enough sustenance to get them through the day. I suspect that their husbands, occasional consorts, or whoever the fathers may have been were no more understanding than the men of today. And eventually, when the child unceremoniously dropped out of the womb, the caretaking duties fell upon the lactating female, who had to feed and placate the insistent and unforgiving baby. It could not survive otherwise.

There are many aspects of human infant behavior that would seem to be less than ideal. First is the complete and utter dependency of our newborns. They are totally useless blobs that do little but cry, drink, and excrete. The altricial young cannot be left alone, for they constantly put themselves in danger. We can be grateful for the evolution of a human nature that finds these creatures cute, so that infants can generally elicit the high degree of care they need. Natural selection could not have worked any other way, for if we did not care for our dependent offspring, our numbers would quickly dwindle to zero.

Many other animals have dependent young as well, although not normally for such an extended period of time. An infant baboon is an uncoordinated tangle of four legs that seems to have an insatiable, humanlike curiosity about everything. Fortunately, when it tries to crawl away toward an object or activity of interest, the mother can just nonchalantly grab its skinny tail, treating it like a natural leash. And when it is time to go, mother just scoops up the infant to her underbelly or sometimes her back, where it grabs on for dear life, clutching mightily to her hair while she gallivants up and down cliff faces or across the plains.

Human babies, and human parents, are not so well equipped with the accouterments of survival and transport. There is no tail to grab, and when the infant starts to crawl away, the mother must gather up the whole messy bundle of a child into her arms. Our early evolutionary loss of the tail (apes and *Australopithecus* did not have tails either) may have deprived predators of one more appendage by which to catch us, but parents lost the natural leash to their children.

Transportation across the savanna or through the forest was a matter for which human families with babies or toddlers were at a distinct disadvantage. Today we have all sorts of contraptions to help us tote our burdensome young, but in the early days I am sure that mothers wished they had a natural kangaroolike pouch into which they could shove the wriggling child. There is no fossil record of when we lost our body hair, but for some time there has been no chance of a baby clinging to the mother's body like the living baboons at Taung and Makapansgat.

There are many theories about why our ancestors became bipedal. One popular theory, which has held up to close scrutiny over the years since it was proposed by Darwin,[3] is that by becoming bipedal we gained free arms and hands. These could then be used to carry things, make tools, and so on. I suspect, however, that our upper-limb advantage has long been used for carrying children, even though the little tykes can get quite cumbersome on a long afternoon walk under the African sun. It is yet another evolutionary compromise—the advantage of arms being lost at times to the disadvantage of carrying children.

Demanding Children

I should like to dwell on the topic of infants for a while longer. As father of two, I find it to be of particular interest. Before each cherished, albeit difficult moment of birth, while the fetus in my wife's womb grew, slowly disabling her formerly graceful bipedal gait, we knew what to expect: sleepless nights appeasing our loudly crying dependent.

Crying is a language of its own. Loud shrill cries, persistent wailing cries, whimpering babbling cries—some claim to be able to interpret them in terms of discomfort, desire for food, and so on. But these piercing noises still wake up the neighbors. In a more natural environment, the high-decibel cacophony of modern little human cherubs certainly would have alerted hungry predators three valleys away. The unanimous interpretation of the all these predators' small but focused carnivorous brains: mealtime.

How did our ancestors ever survive when every saber-toothed cat, lion, cheetah, leopard, hyena, and jackal within earshot was alerted to the presence of a group of slow bipeds carrying bellowing babies in their tired arms? At Taung and Makapansgat I have spent many nights camping out near troops of baboons. The infants are virtually silent. Occasionally two young juveniles might have a raucous nocturnal spat, an apparent bout of sibling rivalry, but one sharp bark from the dominant male and all remains peaceful until morning. Baboons have mastered the art of remaining quiet. To be sure, as I discovered on the grisly night of the leopard attack in the Makapansgat valley, they are still vulnerable at night. So why would an early hominid risk even greater predation with loud cries?

In a natural world, the human infant's proclivity for selfishly and loudly drawing attention to its impatient needs would have betrayed the entire group to multiple dangers. But humans are not alone among animals in possessing such perilously vociferous behavior. When I lived in Johannesburg, there was a pair of recently hatched pigeons nestled among the pipes outside my bathroom window. I was alerted to their presence by the shrill squeaks they emitted to draw their mother's attention.

From the upper floor of the apartment building, I often heard adult pigeons outside my windows engaging in their rather riotous reproductive ventures; but at least they could fly away when I cast my irritated glance out the window. The products of their copulatory persistence were entirely more at risk, at least until they fledged and stopped crying out for the attentions of maternal pigeons and potential consumers alike. They were lucky that my interest in them was purely academic and that I did not develop a taste for pigeon chick: I could easily have reached my hand out the window to snatch them both. In the wild, the inaccessibility of pigeons' nests on cliffs makes it hard for predators to get at the chicks, and the chicks can make all the noise they want; the urban approximation that pigeons use for nesting does not necessarily offer the same protection. But, to judge from the number of pigeons in Johannesburg, the briefly perilous state of newly hatched chicks has not greatly hampered the success of the species.

There are ample explanations for why evolution allows such seemingly disadvantageous behavior among the young of many species. Richard Dawkins, in *The Selfish Gene*,[4] wonderfully elucidates a cogent one. In a nutshell the child, the chick, or the young of just about any species that depends upon maternal care for food gains an advantage by calling attention to its hunger. It is rewarded by being fed, perhaps to the detriment of its less demanding sibling, or perhaps to the detriment of the whole group if the mother does not shut it up before it alerts a lurking predator. If the young one is successful in eliciting feeding behavior or protective behavior from its parents, then it is more likely to survive. If it survives, it is likely to pass on the genes that prompted its obnoxiously selfish behavior.

Many complications can arise with such a daring ploy. If parental appeasement fails to quiet the offspring and a suitably hungry predator is attracted, natural selection will very quickly make sure that such genes do not get passed on. I once encountered a woman who professed her horribly crass attitude toward infants with the lyrical sentiment: "Baby cries, baby dies." A few thousand years ago this statement may have had a ring of truth for our ancestors.

The principle of infant selfishness can take more severe modalities than human or pigeon vocalizations. Black eagles, like many other raptors, start their lives in a most peculiar fashion. At Makapansgat I reveled in watching a magnificent pair of adult black eagles, who mate for life, soar high above the valley after taking off from their aerie on the red cliffs. I recall leaning over the top of the cliff one afternoon, looking down on the distinctive white V on each bird's back as they caught a thermal and rose effortlessly in front of me. It seemed that the lovely pair had a blissful life of ease, gliding over the abundant prey below. But their serene and seemingly glorious lives begin with horrific sibling competition and treachery that dwarfs the benignly obnoxious behavior of youthful human brothers and sisters.

Black eagles, like the pigeons outside my Johannesburg window, are normally born in pairs. Soon after hatching, the eagle chicks invariably fight to the death of the weaker. The victor consumes the very flesh of its sibling, unaware of the cannibalistic nature inherent in its consumptive drive. Quite independently, spotted hyenas have also evolved a propensity to attack and often kill a twin sibling cub. It can be a vicious world among carnivorous beasts, and it is sometimes difficult to imagine the long-term evolutionary benefits in light of such manifestly despicable behavior. Survival of the fittest indeed.

Mind you, in some past human societies, should twins be born, the evil mother and her two offspring were *all* killed!

Incidentally, more benign forms of competition can also take their toll. The duet of pigeon chick voices chiming outside my window was soon reduced to a solo. The more persistent chick had eagerly nabbed food from the mother's crop while the other gradually became quieter and more uninterested. Getting weak from lack of nutrition, the quieter chick succumbed to internal "predators," perhaps viral or bacterial, and was hushed forever. One might think that this outcome would favor the presumably superior genes of the surviving chick, as in the more brutal competitions between black eagle or hyena siblings, but three days later the winning pigeon chick also lost its life. Even without the swift and fatal pounce of an external predator, a chick's, cub's, or baby's life is as fragile as it is precious.

The point is that the evolutionary benefit of outcompeting a sibling often goes to the individual, not necessarily to the group or the species. It is a tradeoff, a compromise in which the need to cry out and be fed outweighs the vulnerability of being fed upon. As long as enough individuals survive to maintain the species, there is no reason why such a competitive system should not evolve; in some cases, especially among carnivores, such early rivalry and pruning of the young population may prevent later competition for a limited food resource. Often enough, this strategy works. The spotted hyena (the species *Crocuta crocuta*) has survived in southern Africa for at least 3 million years, as evidenced by fossils at Makapansgat and later sites. We do not know if their twin cubs always attacked each other, but for however long they have done so it seems not to have hindered their evolutionary success. Three million years is a long time for any species to survive (the average large mammal species in southern Africa appears to have survived for roughly 700,000 years before evolving into something recognizably different or going terminally extinct), and the numbers of spotted hyenas are still reasonably high despite human encroachment on their habitat. On the other hand black eagles are having a greater difficulty sustaining their populations in the modern world.

I sometimes wonder, however, if such an evolutionary peculiarity as infant selfishness, mildly displayed by the harsh cries of human infants, severely limited the numbers of the early humans who left us our legacy. Behavior does not fossilize, but if loud crying has deep evolutionary roots, it may have been an important factor in attracting unwanted predation and pruning the population. The growth and decline of populations must be an important consideration for the evolutionary scientist, as we found with the fickleness of population trends in the computer simulations. If natural selection promotes a feature that puts population numbers in jeopardy, it would take considerable good fortune for such a population to survive. Or it would take an evolutionary tradeoff to counterbalance the ill effects of risky infant behavior.

Speaking of Which . . .

Human communication—vocalizations beyond the cries of our boisterous infants—has been another key factor in our evolution, involving advantages as well as consequences. Again, it is difficult to trace the origins of speech, but we do have clues. South African paleoanthropologist Phillip Tobias, carrying on the tradition of his predecessor Raymond Dart, has taken great interest in the evolution of the human brain. Cranial endocasts, like that of the Taung child, can reveal an amazing amount of anatomy reflecting the convolutions and characteristics of ancient brains. Such endocasts of *Homo habilis,* Tobias discovered, reveal subtle enlargements of the cerebrum at the specific brain areas that are known to govern speech and word association.[5] Whether or not these early ancestors of ours actually spoke as we do, with complex sounds and subtle intonations, must have depended on the anatomy of their larynx.

The larynx, through which air passes on its way into and out of the lungs, is an amazing complex of soft tissue, ligaments, and cartilage. These components work with one another to tighten the vocal cords or narrow the gap between the cords in order to alter the pitch and character of sound, produced as air passes through the gap. One such important component is the thyroid cartilage, often referred to as the Adam's apple. Unfortunately, no cartilage or soft tissue gets preserved in the fossil record, so we are left to deduce indirectly when and how this structure evolved. One key component of our speech-making ability is the low position of the larynx in the throat, as you can see and feel on yourself: the larynx is just behind the prominent thyroid cartilage, well below the mouth.

The low position of the larynx is unique among humans, and it seems to be related to our highly flexed cranial base.[6] Raymond Dart first recognized the human affinities of the Taung child by the position of the foramen magnum: this large hole at the base of the skull was quite far forward on the cranial base, and thus not only closer to the center of gravity (making it suitable for an upright stance) but also closer to the bones of the face. Its position continued to change throughout human evolution. In between the facial bones and the foramen magnum there is a lot of important anatomy, all scrunched together. In most mammals this area of the cranial base is more spacious and occupied in part by the laryngeal apparatus. In humans, however, our upright posture pushed the face and the spinal cord closer together, resulting in an increased flexure at the base of the cranium. In the wake of these changes, the larynx had to get out of the way. It could not go up, it could not go sideways, so it had to go down.

The evolutionary descent of the larynx had a lot to do with the successful descent of modern humans. As the larynx dropped lower, a chamber was created through which the voice could resonate and be manipulated by articulate movements of the mouth and tongue. This gave us the full range of vowel sounds we use so effectively today. One may assume that following the time of *Australopithecus*, as early *Homo* coupled the neural mechanism of speech with an increasingly flexed cranium and lowered larynx, the incipient ability of humans to speak was continuously enriched.[7] But at a price.

In other animals the larynx is positioned directly behind the oral cavity, and the opening to the larynx is up nearer the back of the nasal cavity. So, for example, when your pet dog greedily devours its meal, the food passes around the soft tube of the larynx, safely below the laryngeal opening that allows entry to the airway. You cannot wolf down your food with such impunity. The low position of your larynx means that food has to pass right over the opening to the larynx. One false move and you may choke.

A number of mechanisms have evolved to counter the human susceptibility to choking. Gently place your hand over your throat as you swallow, and you will feel your thyroid cartilage, the Adam's apple, rise and fall. Your entire larynx is lifted each time you swallow so that the laryngeal opening is pressed against another cartilage, the epiglottis, which is pushed back by the tongue. Other tiny muscles help to narrow the opening of the airway and to bring the epiglottis down over the hole through which you try to breathe. But the whole complicated mechanism is inadequate, if not a bit of a mess. The two main things we must do to survive, eat and breathe, have become precariously mixed, and often with fatal consequences. If you do not believe me, then try to breathe when you swallow—but I strongly recommend against such an experiment. Our laryngeal scar of evolution will not allow it.

Taking Evolution into Our Own Hands

We have paid costly dues for whatever selective advantages are afforded by our bipedalism, large brains, and vocal capabilities. But we got what we needed to survive. And we seem to be doing quite well from an evolutionary perspective, for there are many of us who survive across the globe—over 6 billion of us. That's a lot of bad backs and hemorrhoids. Any one of us may choke to death any minute.

Prodigious brains and a refined linguistic apparatus have carried us far, as have our two callused and overburdened feet. Culture, the natural by-product of our cerebrations and communications, has been good for our species. It allowed us to overcome the compromises to which our bodies were subjected. The wounds inflicted by our evolutionary heritage were eased into relatively unobtrusive scars. Now we can adequately protect the vulnerable child or even the strained youth, as well as the limping aged. When necessary, we may forcibly expel the food from a choking victim's throat and learn from the experience.

Although we are inadequately adapted in some respects, perhaps even maladapted, our overall fitness comes in a different form. Human fitness comes from adaptability. Culture, learning, and indeed foresight make us adaptable to almost any situation. This can make up for conditions where biology fails us. For human minds, and human minds alone, necessity can be the mother of invention.

But just as culture has allowed us to overcome some of our biological deficiencies—the scars of human evolution—it may be creating still more deficiencies. Our evolution and/or "devolution," as some would mistakenly call it, continues today. The cultural environment in which we now live redefines the selective value of biological features. This generally works more toward the reduction of features than toward the incorporation of new features. We lose more than we gain.

The relative lack of acquisitions is easy to understand. For example, although fathers have taken on an increasingly important role in the day-to-day rearing of a child, at least in modern Western society, men cannot expect to start lactating to feed our offspring. Men may have two nipples, a point of great consternation for Charles Darwin, but we men are missing the hormonal structure that grows the proper glands and produces milk. It is not impossible that someday an otherwise normal male will be born with a peculiar new combination of genes that allows lactation, but it is doubtful. It would be a chance development of low probability, and it would have to chance upon a suitable environment before it would be of selective value. Men who may want to feed newborns cannot wait for biological evolution to comply, and so artificial bottles must do. That in and of itself changes the rules of natural selection.

The artificiality of a human-dominated planet changes the rules of natural selection for many animals besides humans. Pigeons, for example, seem to thrive in city environments. The buildings on which they roost, as well as the statues on which they tend to answer the call of nature, are not very different from the rocky outcrops that are their natural homes. So they have not adapted to city life, but fortuitously they were preadapted—for part of urban life. The mangled pigeons found on roads every day indicate that they have not entirely adapted to their new homes and the unnatural vehicles that traverse their feeding grounds. One might think that there would be strong natural selection for pigeons that get out of the way of large, shiny, fast-moving objects. But selection must have variants from which to select. If the genes that hardwire the limited neural circuitry of the pigeons' little birdbrains do not chance upon an appropriate perception and response mechanism, then pigeons may litter our roads forever. No species gets quite the features it might want.

The goats at Taung have a similar problem. Their evolved instinct is to stay with the herd. When a car comes barreling down a dirt road that goats are crossing, kid goats that have not yet crossed will frequently risk their lives by dashing in front of the car. Such behavior may have made sense at one time, but it seems selectively foolhardy now. Goats, like pigeons, are stuck with their maladaptive behavior in the face of encroaching civilization and have little hope that things will change. But their populations are doing all right, at least for now. They are adaptable enough to human culture.

For all the disturbances of nature that we humans inflict upon the planet, the most interesting consequences may be the ones that befall our own species, *Homo sapiens.* The tools and products conceived by our savvy brains and constructed by our free, adroit hands supplant the biological tools with which nature supplied us. It often does not matter that we cannot acquire traits we may want. With warm clothes, for instance, it does not matter that people who live in cold climates never acquired natural fur coats. But in such situations a general principle applies to the continuing evolution of features among mammals: if you don't use it, you lose it.

There are many examples of formerly functional features being lost, due at least in part, perhaps, to cultural effects—and lack of use. The most obvious come from modern medicine. Medical prowess has extended our lives well beyond the forty years or so previously allowed by the natural world and accommodated by natural selection. After the limited life span for which evolution has honed our bodies, medical doctors prop up our degenerating bits and pieces. But one need not get old before medical technology starts changing our features.

Many of you will wear eyeglasses to read this book or rely on less obtrusive contact lenses. How well do you think you would have survived with poor eyesight just a few thousand years ago? Reading would not have been necessary back then, so the effect of poor eyesight on your life would have depended on the severeness of the problem. And your eyesight may not have been so bad if you were born thousands of years ago. Indeed reading, staring at computer screens, and other potentially eyestraining activities required by our magnificent culture may be causing some of the deficient vision of modern times. Yet many people have visual inadequacies born of the genes they inherited from their parents. And before the advent of corrective technology, they would have entered the struggle for survival at a distinct disadvantage.

The visually impaired of long ago lived in an unforgiving world. Just imagine being in their shoes (or the lack thereof). You would have been a very ineffective hunter, unable to take proper aim at your prey, or an inefficient gatherer. You and your family would have been hungrier than the rest, for even the most successful and magnanimous humans would probably have chosen to feed their own family members first. All this assumes that you would have lived long enough to have a family. With poor eyesight you may not have recognized that the animal walking toward you was not your faithful hunting dog but a hungry leopard.

Today your imagined story is likely to be quite different. Although not all have the opportunity to benefit from innovations such as eyeglasses, the spread of such cultural amenities still astounds me. I find glasses appearing even on the proud faces of the most impoverished people at Buxton. Yet any population geneticist can tell you that hidden effects must be spreading as well: genetically coded "defects"—and I use the term with some hesitance—are being allowed to accumulate in our species. Everywhere.

Because medicine has changed the rules of natural selection (as have the use of shelter, clothing, and many other ingenious products of our busy cerebral activities), new genetic variants, such as those coding for poorer eyesight, can accumulate. And because most new variants tend not to be helpful, we increase the "load" of seemingly maladaptive genes within the population—which is not necessarily bad, because what is adaptive or maladaptive depends on the environment. One person's supposed genetic defect may be another's benefit somewhere else or at some other time. But however we view the results, there is no escaping the fact that we have created our own environment, defined our own ecological niche, shaped our own selective forces. Our evolutionary successes have catalyzed our culture, which in turn creates new environmental contingencies (of our own making) for further evolution. And that is autocatalysis writ large.

If a new beneficial variant were to come along—and I have no idea what that would be—natural selection would still run its course and perchance, with some luck and time, would establish the new feature in the population. It is difficult to speculate about what selective forces may be acting upon our species today, and it is too early in this book to hazard a guess. However, it does seem clear that in the meantime we will continue to reshape our bodies with the loss of structures we do not use. Useless or unnecessary parts of our anatomy will become vestiges, like the appendix, or disappear altogether. This will not occur because of natural selection but because of the *relaxation* of natural selection.

The notion of use it or lose it may at first seem ludicrous. How can a directional change, such as the loss of structure, occur without a directional evolutionary force such as natural selection? Surely chance mutations will arise that affect the structure in question, but chance alone cannot sustain a directional trend. Or can it? When random genetic mutations accumulate, under the relative indifference of natural selection, we get what is known as the "probable mutation effect." The normal, most probable cumulative effect is structural loss—the reduction of a feature or of its ability to function—and our eyes may prove to be an example.

The idea of probable reduction was conceived by the ever eccentric anthropologist Loring Brace. I consider Brace to be my academic grandfather, for he was the mentor of my mentor in the American university system. Yet despite such proud links I initially found his notion of a random reduction mechanism to be rather unlikely if not unfathomable. But his logic works, and so do the numbers needed to support the hypothetical effect, as I found out in my first foray into computer simulations.[8]

Brace's hypothesis is actually quite simple. Genetic mutations, we have seen, are fairly commonplace. New mutant alleles that turn out to code for useful features may survive, whereas new alleles that reduce the fitness of an organism are normally eliminated with time. One can assume that our human genes are fairly highly selected after all these years, as are the genes of pigeons and hyenas. Thus an error in copying a genetic instruction is not likely to benefit a living organism. To use an example from the previous pages, an error in the coding of genes for lactation is more likely to make a female lose her ability to produce milk than it is to make a male start lactating. Both occurrences may be possible, but the negative effect is far more *probable*—things are more likely to go wrong. (Murphy's Law strikes again.)

Likewise, although the four-chambered heart has been pounding in all kinds of vertebrate species for some 200 million years, mutations have yet to come up with a significant improvement (that we know of). But mutations that affect heart structure to its detriment keep arising. And in general the *most probable* mutations are those that produce negative effects: they result in a loss of function or reduction in the size of a structure. It is inherent in the genetic algorithm of life. It is inherent, in fact, in the universe: it is entropy.

So mutations are always occurring, and they tend to lead to structural or functional reduction. However, if the loss of the feature in question entails no survival disadvantage, as would be the case for a structure no longer used much or needed, then natural selection will not weed out the new allele coding for its reduction. The allele will stay, and many more such mutations may accumulate in the population. Entropy thus becomes evolution.

Let's start with a nonbiological example to illustrate the principle of the probable mutation effect, or PME. As I write this book on my computer, typographical errors constantly creep in. I am a writer, not a typist. Fortunately the word processor on my computer has a subprogram to check my spelling. Let us assume that the spelling checker is the selective force in the world of properly constructed manuscripts. Some words, such as *there, their,* and *they're,* may all come out as equally viable options: the selective force may not notice that there usage at that specific point their in the manuscript does not reflect they're proper function as words of the English language. If the sole (not soul) selective force is the spelling checker without a context editor, then the errors may slip through. There is no lack of selection, for the misspelling *ther* constantly irritates my machine and spurs it into action. However, the relaxation of selection from proper English standards to standards recognizable by a computer has resulted in structures that are functionally reduced. Mutations that used to be weeded out now become viable alternatives, and this allows "mistakes" to creep through. Believe me, they accumulate with time and sometimes go undetected until after publication.

As with the PME in biology, certain kinds of errors may accumulate in manuscripts, depending on the selective environment. Such analogies are instruction, butt are knot necessarily worth there wait in papier.[9] One problem in this analogy is the implication of a purposeful outside force, such as a reader or an editor, determining the value of alternative spellings. Likewise in evolutionary theory, *selection* tends to imply a selector, hence our problem with Darwin's choice of the word. In biology there is no outside force imposed directly upon the genes of a plant or animal. There is no editor of life. If an animal endures the rigors of nature, or a book survives the market forces, then either may be reproduced over and over again by the passing on of genes or the rolling of printing presses, with or without mutations. It is survival of the survivor. But now that the principle of the PME has been roughly established by analogy, we can take a more rigorous look at some biological examples.

The human eye is a highly complex structure built and controlled by many genes coding for many proteins. All the resulting features have undergone the metaphorical scrutiny of natural selection for a considerable time. A flaw in just about any of the genetic instructions may result in component parts not fitting or working together in precisely the way that leads to optimal vision. The chemistry may be wrong for transforming light into a nerve impulse received by the brain, or the shape of the eyeball may be wrong for the focal length of the lens. Such defects can lead to slightly impaired vision, severely limited vision, or even total blindness. In most of the milder cases eyeglasses, surgery, or other corrective actions provided by our adaptable human culture allow the individual to live a totally normal life. He or she has no exceptional difficulty in passing the defective gene to the next generation. Such genes then accumulate like typographical errors going into book after book.

I do not suspect that humans will necessarily lose their eyesight altogether, for vision still plays a very important role in our existence, with an undoubted survival (and hence selective) value. In this case as in many others, the probable mutations are taking effect only with relaxed selection, not an absence of selection. Natural selection will likely maintain a base level of functioning for the average person, although our collective eyesight may be somewhat reduced from the visual acuity of the human race as it was before medical intervention altered the selective environment. But if human vision were to become necessary no longer, for reasons of culture or otherwise, we could inherit a fate like that of the blind crickets and fish that inhabit deep, dark caves across the globe.

Life Deep within the Caves

Charles Darwin was intrigued if not perturbed by the biology of cave-dwelling creatures. I refer not to our somewhat mythically troglodytic "cave men" ancestors, who really spent most of their time under the sun, but to the enchanting beings that permanently reside in the darkest reaches of caves. Such animals, particularly crickets and small fishes, tend to lose features related to seeing or being seen, such as eyesight or coloring, even when those features have a high selective value for their close relatives still living in a world of sunlight.

Many cave animals, or troglobites, are totally blind. Their eyes are nonfunctional vestiges of the complex structures that translate light into messages for the brain to interpret. But of course there is no light to perceive deep within a cave. So there would be no use in evolving better eyes for gathering limited light, as many nocturnal animals have done. And without a use for eyesight, there is no selective advantage in maintaining a functional visual system.

Likewise, cave fish and crickets have lost their coloring. Most are fairly bland, somewhat translucent creatures. Deep within a cave, devoid of even the smallest amount of light, the ability to identify a creature by its colors matters to neither potential mates nor potential predators. So body coloring, as useless in the dark as eyesight, has been lost among most troglobytic creatures.

What is most interesting about cave creatures is that their populations often evolve independently of one another. A set of chance circumstances may bring a few founders or their eggs into a cave at some time, but then the population usually becomes trapped and totally isolated from the outside world, as well as from the worlds of other cave creatures. So time and time again, in unique populations sequestered among separate caves, the same evolutionary trends have occurred.

Be it a fish or cricket or another creature, each tends to lose sight and coloring. This suggests an evolutionary principle at work.

Recognizing an evolutionary trend is easy; discerning the principle at work is not necessarily so simple. For each case of blind, colorless cave animals there are at least two possible explanations: natural selection for structural reduction, or the *relaxation* of selection leading to the PME (the same result by a different process).

The theoretical perspective most often assumed is that natural selection favored those cave-dwelling individuals without eyes or color. But what would be the selective *advantage* of the loss of eyesight? Some biologists suggest that there is an energetic cost in producing and maintaining an eye or coloration. If the eye is not needed, then energy that could have gone to the development and maintenance of other structures would be wasted on an eye.

This energetic cost hypothesis is conceivably valid and testable, but it is still waiting to be thoroughly tested. On the other hand one may suspect that natural selection is not so efficient as to recognize minor developmental costs. We have already seen how natural selection cannot possibly cope with every scientifically conceivable gain potentially provided by each and every gene. Its progress bogs down in the complexity. Furthermore, the inefficiency of natural selection is especially apparent in small populations, such as those that inhabit and exploit the little that is available for sustaining life within the deep reaches of caves. Natural selection has tremendous power, but it is not ubiquitous and applicable to every case of evolutionary change. Even Charles Darwin had difficulty envisioning natural selection alone as the cause of structural losses among cave creatures. He attributed the peculiar features of cave dwellers to the effects of disuse.

Darwin, you must recall, had not been enlightened by a knowledge of genes and genetic mutations. He could not have conceived of the probable mutation effect, although his insight into nature almost foreshadowed Brace's PME, much as it almost foreshadowed Dart's discovery of early hominids in Africa. Darwin had many prescient insights.

The PME hypothesis, resting on the shoulders of genetic theory, is a feasible and practical alternative to natural selection when it comes to explaining evolution in caves. With the relaxation of natural selection, things still happen to evolving species, but they happen in a different way. Either way, with or without natural selection, the gain or loss of a structure still constitutes evolutionary change. But we must look at the logic of the alternative models to assess which is more practicable to explain any given case in nature.

Let us imagine, for a start, a small population of fish blessed with the advantages of vision. Their eggs may have been brought into a cave by chance on the feet of a visiting bird. In the open air, such hitchhiking is not uncommon. Louis Bromfield, a novelist and naturalist from my home county in Ohio, wrote of how a pond he created, dug into the earth near an underground spring, soon accumulated fish populations without artificial stocking. The same or similar phenomena might happen, albeit rarely, to initiate an isolated population in a cave pool. Alternatively, a stream leading into a cave from outside might dry up following a climatic change or the tectonic uplift of a mountain range, leaving a cave pool isolated but full of life. And water itself can find many ways to get into a cave, bearing all kinds of debris, not to mention what the leopards might drag in—as we learned when searching for relatives of the Taung child. What happens once a creature is there, living or dead, is left to the whims of circumstance. Wrote Bromfield: "Those who live near water know that in the business of carrying on life, nature can be incredibly tough and resistant and overwhelming."[10]

So now, in the shallow waters deep within a cave is a fish population that has managed to survive. There is no light for photosynthesizing plants, and hence no plant tissue at the bottom of a normal food chain. The fish find sustenance instead on the organic benefits of bat guano or other such natural offerings. It seems an unlikely way to live, but in cave after cave we find that life goes on, despite the odds against it. Having found a home and sustenance, the cave's inhabitants no longer require vision, as it is not particularly useful in the dark.

If all the fish in the initial population have the biological potential to see, albeit impossible now without light, then it would take genetic mutations to alter that genetically controlled capacity. The most probable mutations would be those reducing vision rather than, for example, allowing vision in the infrared spectrum, which might have advantages in cave environments. Infrared sensors in video cameras allow humans to see in the dark—something a cave fish might "want" as well to see food. But of course genetic mutations reducing the function of sight are more likely. This much of the PME hypothesis is not at issue; our concern is what happens next. Such reductive mutations could mean that the organism expends less energy developing an eye, increasing its chance of survival. Alternatively, those very same mutations might simply accumulate with time, offering no significant advantage or disadvantage. The latter possibility, the PME in action, is the simpler model. The loss of sight arises from random mutation alone and requires no additional selective mechanism. Darwin put it this way: "As it is difficult to imagine that eyes, though useless, could be in any way injurious to animals living in darkness, I attribute their loss wholly to disuse."[11]

Loss of color is just as easy to explain without invoking natural selection. It cannot be difficult, genetically speaking, to lose pigmentation. I hesitate to bring up this example again, but those wondrous cockroaches of mine have taught me a lesson with regard to cave animals. Having had the misfortune of seeing a mass of newborn cockroaches on one occasion, I know that they are without pigmentation at birth, looking somewhat translucent like cave crickets. The genes that control the subsequent development of the pigment that darkens their color must be fairly simple, and easily rendered nonfunctional. Even adult roaches, which shed their dark exteriors in a molting process, can be translucent for a few hours. If there were no selective advantage to maintaining pigmentation, there is little doubt that eventually cockroaches could lose their color through the probable mutation effect.

What can we learn from the principles underlying the PME? First of all, natural selection must constantly weed out mutant alleles just to maintain certain features. Therefore, as the limited power of natural selection gets busy on genes for new features, it becomes more difficult for a species (or subpopulation) to keep existing features.

The other important point emerging from this discussion is that our cultural behavior plays a very important role in human evolution. As culture takes over various biological functions, parts of our anatomy and physiology become redundant, and eventually we may lose such features. In other words our own behavior blindly sets the course of evolution and initiates new rules of natural selection.

Acquired Characteristics

One feature that human populations in northern latitudes have lost is much of the pigmentation in the outer layer of their skin. Not unlike cave animals losing their color in the absence of light, some of us have lost the melanin content of our epidermis, along with the protection it provided from the ultraviolet radiation of the sun. There may have been valuable adaptive consequences to such a loss. For example, those of us with lighter skin can more easily absorb the ultraviolet radiation necessary for the synthesis of Vitamin D in our bodies; otherwise we would have to ensure that our diets contained sufficient quantities of this vitamin that is so important for building and maintaining good bones.

Fortunately for inherently pale-skinned people who now reside in places with intense solar radiation, such as South Africa, the human skin can usually respond to radiation by producing more melanin. I can acquire quite a splendid tan even in the winter at Taung (although pink or red seems to be my color in the summer). That tan, however, cannot get coded into my genes and passed on to my children. Any trait I acquire during my lifetime by interaction with the environment cannot be inherited by the next generation.

Some time ago biologists believed that acquired characteristics could be inherited. Such a notion is often attributed to, or blamed upon, Jean-Baptiste de Lamarck, one of the great scientists of the early nineteenth century.

Jean-Baptiste Pierre Antoine de Monet, Chevalier de Lamarck, more commonly referred to by shorter appellations, was a French naturalist of considerable note. As the professor of insects and worms at the Muséum National d'Histoire Naturelle, Paris, Lamarck became a great taxonomist, ordering life as he saw it on a scale from the simple to the complex. By 1800 Lamarck had seen through the dogma of the day that species were fixed and immutable entities, and so he began developing the concept of "transformism," or the transformation of species.[12] Long before Darwin, Lamarck was developing a theory of evolution.

Almost everywhere biology advanced, it seems that Lamarck was involved. Long before Thomas Huxley wrote his influential *Man's Place in Nature*, in which he argued for the evolution of humans by comparing their anatomy with that of the great apes, Lamarck had written about the similarities between orangutan and human biology. Indeed we noted earlier that Lamarck had his own prescient insight in 1809 regarding the evolution of bipedalism before brain expansion among ancestral humans, and he even gave us the word *biology*. So we owe him a considerable debt of gratitude.

Despite his great accomplishments and insights, Lamarck gets a lot of bad press in the biological literature. Unfortunately, he never conceived of natural selection as the driving force behind his transformism, or evolution, of species. Instead he argued for the inheritance of acquired characteristics, as many did in his day. Under this principle my children should be able to inherit at birth the tan I acquired under the African sun. The idea of evolution through the inheritance of acquired characteristics has become known as Lamarckism, but giving his name only to that part of the theory focuses on the errant mechanism he perceived rather than the evolutionary concepts he also conceived.

Today the words *Lamarckism* or *Lamarckian* are used almost as pejoratives. Indeed, I have been accused of being Lamarckian for my support of the probable mutation effect. But I too disagree with the mechanism behind Lamarck's transformism. I do not believe that acquired traits can be passed to offspring; I do not believe you can inherit your parent's suntan. Thus, with all due respect to Jean-Baptiste de Lamarck, I should like to focus for the next few pages on the errors of Lamarckism and the triumph of Darwinism. It illustrates an important point regarding *why* you can't always get what you want.

Darwin himself may have slipped into a touch of Lamarckism, although he would not admit it, with his assessment of the effects of disuse. But there is a distinct difference between Lamarck's ideas and those of Darwinists such as myself. That difference was first pointed out just before the turn of the twentieth century by the great German scholar August Weismann.

Weismann had a theory about the mechanisms of heredity quite distinct from Lamarckian inheritance of acquired characteristics. He imagined a substance in all living beings that somehow carries inherited traits—a substance he called the "germ plasm." His germ-plasm theory was an early forerunner of genetics. Because nothing was known of genes in Darwin's day, the origin and transmission of varying traits, on which natural selection supposedly acted, could not be explained. But Weismann saw quite clearly that something more discrete was at work than the Lamarckian passing of acquired characteristics, for the experiences of a lifetime did not seem to touch the germ plasm.

Weismann conducted crude but convincing experiments to prove his basic point, albeit not without trauma to some unfortunate caged mice in his laboratory. Quite simply, he cut the tails off the mice and let them go about their business. When the tailless mice reproduced, their offspring had tails; they had not inherited the acquired characteristic of taillessness. Weismann continued this gruesome task for nineteen generations of unfortunate mice, and still baby mice were born with complete tails.[13] He really should not have been surprised; after many centuries of circumcision among innumerable human males, little boys are still usually born with their foreskins intact.

Weismann thus knew that something deep within the animals' biology, something more subtle than he could discern in gross anatomy or even under a microscope, controlled the inheritance of characteristics: it was the germ plasm to him, genes to us. He also recognized that the series of steps between the transmission of germ plasm and development of the final product—from fertilized egg through fetus to infant, child, and adult—was very complicated. One misstep along the way and the construction might not proceed efficiently. In other words Weismann recognized probable mutations and their effects, or the PME, although he did not have the modern terminology for it. And so, like Darwin, he realized that disuse of a feature would relax natural selection and allow that feature to regress.

Thus, *it takes the constant vigilance of natural selection to maintain features,* as well as to ensure the propagation of new features. Without natural selection maintaining them, features may eventually disappear. This can easily give the *appearance* of Lamarckian loss through disuse—as if not using a feature during one's lifetime diminished it, like a muscle that atrophies, and this acquired characteristic then got passed to the next generation. The eventual effect is the same as the probable mutation effect, but the mechanism is very different.

The best-known example of Lamarckism is the Lamarckian explanation of giraffe evolution. According to the theory of transformism through the inheritance of acquired characteristics, the giraffe attained its stature by stretching its neck to reach leaves on the tops of trees. There it was free to browse where potential competitors for food could not reach; height alone was the barrier keeping others out. The stretched neck, a characteristic acquired during life, was then passed on to the next generation of giraffes, and they stretched a bit further. Or so it appeared to the Lamarckians.

In the fossil record such a scenario almost makes sense. Some 3 million years ago, at Makapansgat, there was a short-necked giraffe

browsing in the woods; we now know it as *Sivatherium maurusium,*
a name somewhat less enchanting than the creature deserves. After
this time, the forest began to decline, and the savanna spread. Browsers
would have been under increasing pressure to attain food, and no doubt
few could reach to the treetops. Modern giraffes then begin to appear
in the fossil record in increasing numbers, and with longer necks. It is
difficult for me to describe this sequence without couching it in Darwinian terms, for the idea keeps creeping in that a few giraffes already
had longer necks, and that natural selection promoted this feature
because it gave them a competitive advantage. But Lamarck felt that
the changing African environment forced giraffes to stretch their necks,
and that the need for a new adaptation drove the transformation. He
was wrong, not only because of the chance origins of new biological
features, but also because of the chaotic way in which features become
integrated as a whole body.

Adapting to One's Own Body

It was certainly easier for August Weismann to experiment with the
effects of *disuse* in mice than the effects of *use* among giraffes. Somehow it is difficult to envision Weismann stretching any animal's neck
in his lab, generation after generation, to see the effect on the offspring.
Indeed the complexity involved with stretching giraffe necks illustrates
the flaws in the Lamarckian idea of inheriting acquired characteristics; it also illustrates both the marvels and limits of natural selection.
Chaos is in the works.

One might think that it would have been fairly easy to acquire a
longer neck. Or so it would appear from the fossil record of giraffe
bones. There was no need for additional cervical (neck) vertebrae, for
giraffes, like us, have seven vertebrae in their necks; they just needed
to extend the segments they had. But it was not quite so easy. Earlier
in this chapter I discussed some of the circulatory problems we have as
upright bipeds. Getting blood to flow even higher, through the tallest
creature on earth, is a particularly difficult biological problem. Life at
the top is not easy.

One of my colleagues back at the University of the Witwatersrand,
Graham Mitchell of the Physiology Department, studied giraffe circulation as well as other aspects of the animal's unique biology. He
explained the many difficulties encountered by the long-necked

giraffe.[14] The most obvious problem is getting blood to the head, about two meters above the heart. A simple stretching of the neck is not enough, for blood vessels must be able to constrict in a highly controlled manner to force blood up to the brain. But just the opposite is the case when the giraffe awkwardly lowers its head to drink water; suddenly there is immense gravitational pressure forcing blood to the head, and a unique network of vessels had to evolve to redistribute the pressure, thus avoiding a massive accumulation of blood or rupturing of the vessels. The system is well enough specialized, both anatomically and physiologically, that when the giraffe has satiated its thirst and quickly lifts its head to the usual lofty heights, it does not faint. You or I could probably not withstand such pressure changes without ill effect.

The giraffe thus had to adapt not only to an environment but to itself. Its large size required elaborate modifications of the usual mammalian circulatory system, both above and below the heart, to accommodate increased gravitational pressures. Just imagine the varicose veins you would have if you simply expanded to giraffe size; I do not wish to fathom the hemorrhoids.

Animals that are becoming larger also have to solve new problems of internal heat regulation: the body has a much larger mass, and yet a proper temperature must be maintained throughout, especially in the sensitive brain. Graham Mitchell has also studied thermoregulation in giraffes; he found that the highly evolved nasal structure and blood vessel networks act together to cool the brain. But this cooling apparatus can dissipate only so much heat. The brain is a long way from the rest of the body in such creatures; being smaller it heats up faster, and out on the savanna the African sun can get very hot. One of the giraffe's solutions for adjusting the transfer of heat is spots, or dark patches. Most people believe that the patchy coloring evolved for camouflage, and indeed I often marvel at how I can stare across the African landscape for a considerable time before my eyes focus on a previously unnoticed giraffe. But another benefit resides underneath the dark patches: a rich network of blood vessels can be infused with blood to release heat to the surface as required. Even Lamarck could not have thought of such a clever solution, had he needed to.

Humans, too, had to adapt to their own size and shape. A change in something as basic as posture or size has ramifications for all parts of the body. Evolution by natural selection sometimes provides an animal with what it needs to survive under conditions of a changing environment, but it does so slowly—not necessarily keeping pace with the ever changing world. It has to be slow because an adaptive change in one part of the body, or in the size of the body, means that many other aspects must change in concert. Natural selection does not always have at its disposal the perfect ensemble of traits an animal may want, hence the glitches that remain in our human biology.

Useful new variants cannot simply be acquired as needed; they must chance to exist *before* they are needed or when they are of some use, and then be selected. Darwin thus argued that natural selection, not environmental change, was sufficient to drive evolution. Just as the early giraffe could not simply stretch its neck to reach uneaten leaves on the tops of trees, it was not enough for incipient hominids simply to stand up and trot across the savanna, as if they could acquire an efficient bipedal gait simply by exercising the right muscles and joints. To do so would have been, shall we say, truly Lamarckable.

Environmental change does indeed have effects on animals. Either their adaptations allow them to survive under the new conditions, or they die and go extinct. The short-necked *Sivatherium maurusium*, *Australopithecus robustus*, and millions of other extinct species are proof enough that animals do not always adapt, or that the advantages their adaptations conferred were fleeting.[15] Delayed extinction is an all too real option, as species that chance upon specialized variants later succumb to the oxbow lake effect. Likewise, the "scars" our bodies still bear from human evolution are reminders of the same principle: you can't always get what you want. Adaptation comes when chance and chaos allow it, not necessarily when it is needed.

The biology of giraffes, remarkable though it is, is not perfect either. There would probably be more of them if it were perfect. Our giraffe expert, Graham Mitchell, brought this to my attention in a way that I initially thought showed his naïveté as regards evolutionary theory. His probing took the form of two questions. He first asked why, if eating from the tops of trees was such a grand adaptation, are the young incapable of reaching such a food source? His question is easily answered in terms of benefit to the species. In general, if adults in their reproductive years have an adequate food source, they produce enough young, and the species continues even if some of the little ones die. (Young giraffes, like other browsing mammals, must compete for the food lower down until they reach greater heights.) On the other hand Professor Mitchell's point was well taken, for it had some punch: mortality among giraffes in their first year has been documented to range up to 73 percent.[16] Obviously the youth are not faring well. The giraffid biological system is so finely tuned that alterations in the developmental theme may prove to be fatal. The same may be true of humans and other mammals!

Professor Mitchell's second question concerned why giraffes did not evolve their long necks earlier, in the more distant past. This point smacked of the same ignorant notions of Darwinian theory with which I started this chapter. A giraffe cannot get a long neck at will any more than a snake can start eating grass or a bird can start to photosynthesize. Mitchell's own answer, however, was very Darwinian and not in the least Lamarckian. The reason the neck evolved when it did was not environmental forcing, as Lamarck would have had it, but intrinsic opportunity. Before a giraffe could evolve a long neck—which would have given it a certain advantage, after all, even before the climate changed and the African woodlands dwindled—it had to have in place the physiological mechanisms to allow it to adapt to itself. At the very least, it had to have the potential to compensate for its changes in size and shape. As a nonlinear dynamic system, its own body determined its evolutionary potential.

Giraffids and hominids were evolving in different ways under the same environmental circumstances: the cooling and drying of Africa. Their biological solutions were very different even though both were stretching, so to speak, in the vertical direction. In some ways the giraffe was more successful in chancing upon innovative designs to overcome gravity for blood circulation. In other ways we can be proud of our own distinguished biological heritage, using those large brains on top of our shoulders. Neither solution is perfect. Both solutions involved an opportunistic accumulation of features that permitted each type of animal to survive its changing environment as well as itself, and among the chaos of competing bodily functions each hit upon a different morphological attractor. In both cases the evolution of one part of the body led to potentials for evolution in other parts.

Human evolution, like that of the giraffe or any other living being, was a matter of trying out random new body parts whenever they happened to arrive. The parts manufacturers were clueless. There was nothing any living animal could do to make them respond to its wants, or its changing circumstances. It was happenstance first, circumstance later. It was not, as Lamarck suspected, a matter of necessity being the mother of invention, for invention is largely a matter of chance in Darwin's world.

Do Not Despair

I suppose that this has been a depressing chapter for some of you to read. It appears that our bodies are falling apart, losing previously well-honed features, and have been doing so for many years. It seems we cannot win the game of evolution, and to a certain extent that is true. Worst of all, our young turn out to be the most vulnerable to the whims of natural selection, and that is quite worrisome. But this gloomy endeavor has a purpose, and that is to recognize a few essential points about the evolutionary process as it builds a body from a myriad of genes.

The first point is that evolution involves adaptive compromises that are as dependent on the intrinsic morphological environment as they are on the outside environment. The human body, or that of a giraffe, is a complex web of interconnected parts. A change in one part requires adjustments in other parts; changes in shape or stature, for example, require changes in circulation. Just as in mathematical equations where one cannot maximize all variables at once, there is no guarantee that all body parts can be at their best or even work together in an optimal manner. We cannot have it all, any more than we can have richly varied vocalizations without the risk of choking to death. We're only human.

The second point is that natural selection starts early. From the time that a fertilized egg in the womb starts to divide and build a new being from its unique genetic blueprint, natural selection has begun. The parts must work in concert during development as well as in adulthood. The moment an infant first cries, or fights with its sibling, selection continues. Fortunately, human siblings have taken on a pattern of sibling rivalry that engenders learning rather than annihilation.

The third and more baffling point is that chance alterations of our genes could conceivably lead to directional evolutionary changes, most probably reductive, without the intervention of natural selection. The losses of eyesight and pigmentation provide examples. On the other hand, this may entail compensations. Whereas evolutionary gains may "distract" natural selection from promoting other potentially useful variants, might these evolutionary losses provide opportunities?

JUST AS GENES compete for a role in the evolutionary stakes, so do body parts. They literally stretch and pull on each other to shape a being, including each human being. Moreover, it is not easy to build a body that has to function during construction. Irrespective of outside environmental influences, the body shapes itself to fit *itself*. This is one reason why Huxley stated that "variability is definite." The behavior and perhaps culture of the final product, the "conditions inherent in that which varies" in an animal, also shape and direct the evolutionary process. That is autocatalysis. Certainly this principle is at work among modern humans, whose manipulations of their own environment have catalyzed countless trends in their own evolution. But a more general principle of autocatalysis may underlie all mammalian evolution. I believe that it does, and that autocatalysis can meld the riotous, unsettling principles of chance, coincidence, and chaos with the more familiar force of natural selection.

8

Autocatalysis

Now we have a big problem: finding an explanation of human evolution. If one were to imagine life as it was 5 million years ago, knowing only that a new kind of primate was about to diverge from the others, the prospect of human evolution would appear highly unlikely. The meddlings of chance and chaos with natural selection could have led our early ancestors down just about any evolutionary path. Yet despite the seeming odds, the path was traversed by beings with two legs and a head barely balanced above an upright body.

The only perspective that makes our evolution appear more likely is retrospective; after all, we *do* exist. We can look at what we are today and try to project the phases of our evolution back into the past, filling in the details with clues from the fossil record. But this is difficult as well. Even in retrospect it is not easy to look at a chaotic system and pare it down to its simplest deterministic components or find the subtle initial conditions. And whereas the fossil record gives us wondrous glimpses of our past, its fragmentary nature leaves many questions unanswered.

The challenge and intrigue of explaining human evolution has led to the rise and fall of many ideas. To arrive at one current synthesis, a small cadre of scientists has thrown some highly prized scientific notions out the window, claiming the following: Environmental change does not cause evolutionary progress. Natural selection is severely limited both in its power to promote useful genes and in its freedom to tinker with morphology. Human bodies are not particularly well adapted in many respects, revealing the chance origins of nature's "designs." Chance and chaos, as much as the ever vigilant selective process, made us what we are.

Is this just a curmudgeonly approach, or is there really more to human evolution than our theories have captured so far? There must be more. Despite our faults, humans are remarkable beings in need of an explanation. Whereas no single explanation will suffice, there must be a basic principle that can weave chance and chaos into the fabric of evolution through natural selection.

There *is* more. There is a causal explanation of our origins, rooted oddly enough in the visions of Charles Darwin and Thomas Huxley, that builds upon the themes we've examined in this book. Moreover, its theoretical predictions actually fit the fossil data. The explanation is autocatalysis. If the elements of chance and chaos in our evolution seem disturbing, then autocatalytic evolution will add the final push into despair, for the hypothesis depends also on the third part of the triumvirate: coincidence.

Epiphany

Autocatalytic evolution simply means this: evolution is caused (catalyzed) by itself (auto). It is self-propelled by feedback loops. If this means that most evolutionary change is catalyzed or caused by the inherent nature of a species, then the grand theories of environmental forcing fall away. Evolution would proceed with or without changes in climate or in the plant and animal community with which a species interacts. Evolution is the cause of evolution, and it continues by its chaotic devices. The theory of autocatalytic evolution is painfully simple, horribly mundane, and probably correct.

It seems most unsatisfying to look at the marvelous achievements of evolution, including ourselves, and say nothing more than they caused themselves. But the theory goes deeper than that and holds a lot of explanatory power that has often been lacking from the neo-Darwinian synthesis of evolutionary theory. Autocatalytic evolution

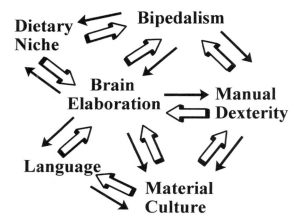

Figure 8.1 • Autocatalytic feedback loop of human evolution. Each morphological aspect has functional or behavioral consequences
or correlates (*solid arrows*), which in turn reinforce the
evolved features through positive natural selection (*open
arrows*). From Poirier and McKee 1999.

is terribly simple and thus tremendously powerful. Moreover, it makes
sense where other explanations of the origin of species fail.

I must confess that when I first grasped the notion of autocatalytic
evolution, I thought it was my own scientific epiphany. But despite
my initial revelation from *seemingly* nowhere, neither the name nor
the concept is new. Autocatalysis is a seldom-used term in evolutionary studies, but the notion has been around since the inception of Darwinism. The term should be used more, and the theory behind it
should be more thoroughly investigated, for a lot of mammalian evolution, perhaps most of evolution, may be a product of autocatalysis.

In human evolutionary studies, the idea of autocatalysis was
spawned in 1965 by the Polish scholar Tadeusz Bielicki.[1] He envisioned positive feedback loops among genetically controlled features,
such as bipedalism and brain structure, and cultural features including tool use and language. Phillip Tobias,[2] with whom I was working
during my alleged epiphany, began to elaborate on the model in 1971.
The brain was the key element in Tobias's model, interacting with
hands, eyes, and particularly the biological mechanisms of speech on
the biological side, all of which interacted with culture (fig. 8.1). The
evolution of the brain allowed manipulative hands, communication

through speech, and cultural transmission of skills from generation to generation. The success of these abilities in making humans more adaptable in turn selected for a larger and more complex brain. In such a model the brain evolved very quickly, starting at the dawn of the genus *Homo* and accelerating by this autocatalytic feedback.

Most models of autocatalysis focus on the acceleration of the pace of evolution, such as that of the human brain. But autocatalysis can be much richer, explaining not only the quickening of evolutionary change but also stasis and indeed loss of features. In light of chaos theory we can extend the model to include genetic, morphological, and behavioral features that are interconnected in an evolutionary system. Even a small change in one part of the chaotic system can have amplified effects through interactions with other parts, for better or for worse. And yet despite such apparent complexity, autocatalytic evolution is a concept so simple, so basic, that it is actually difficult to explain. Intuitively one expects evolution to be much more complex and intricate, hence the elaborate ideas about environmental causation. So first I want to present a scenario of human evolution to illustrate the principles of autocatalytic evolution. After that we shall be prepared to ponder the implications of the hypothesis.

An Evolutionary Fable

What follows is a hypothetical sketch of human evolution. In many respects it is a just-so story, like Rudyard Kipling's fanciful explanations of why things are the way they are.[3] Just as in computer simulations of theoretical models, the following conceptualization is perhaps a grandiose oversimplification of reality. On the other hand, just like a computerized model, it may provide us with an adequate explanation of the overall pattern we see and at least get us on track to find some operative principles.

In this autocatalytic model hominid evolution starts with an arboreal (tree-dwelling) primate as the common ancestor of the lineages that eventually led to chimps, gorillas, and humans. In other words, at some point in our distant past our ancestral stock lived in the trees of the warm, wet forests that covered vast expanses of Africa. We have little idea what these creatures looked like, save for clues from a few fragmentary fossils, but no doubt they consumed a rich smorgasbord of delights offered by the trees. And their populations grew.

It would seem that there was no need for change, but the relentless factory of opportunity hidden within their genes chanced upon an intriguing variant. From the new allele or alleles, the developmental process built a slightly different body. This morphological change allowed the arboreal primate to augment its locomotor and dietary behavior such that members of the species could exploit resources on the ground as well as in the trees. There would be a selective advantage in widening the arboreal niche, especially if resource availability in the trees was limited either by competitors or by the ever growing numbers of the species itself. With more resources the populations could continue to grow. Survival of the generalist, indeed.

If one could get around on the ground in the forest, one could probably traverse the ground outside the forest as well. Thus a consequent effect of the modified morphology would be a "preadaptation" to exploit a niche outside the forest.[4] Out beyond the forest edge there was a lot more ground than tree, with innumerable temptations as well as challenges for survival. Our heroic ancestors may not have ventured very far out of the forest, or very often, but those who managed, courtesy of whatever genetic endowments they had, would have been somewhat freer of the competing crowds that stayed behind. There was no need to rush into new, unknown territories, but the opportunity was there, now that the ground was as inviting as the trees.

In a large population with individuals already shifting back and forth between trees and ground, somebody was bound to hit upon a feature or set of features that allowed a bolder move beyond the forest. Indeed, in at least one population the accumulating advantages of niche expansion led to a somewhat upright posture—allowing, among other things, bipedalism. These emerging hominids could still climb in trees, but they could also walk on the ground, reach food on tree limbs, and perhaps even carry things. Other populations elsewhere, which by chance hit upon a different morphological attractor, acquired mutations that led to the knuckle walking and brachiation (swinging through trees by the arms) that chimpanzees still employ.

There were many *coincidental* by-products of the upright posture, according to this view of early hominid autocatalysis. The most obvious is that it freed the hands, as Darwin noted. Free hands could more easily manipulate objects, reach for food, carry resources long distances, hurl objects in defense. This suite of new behaviors then allowed new evolutionary experiments as hominids began to alter their own course of evolution.

More subtle and intriguing advantages accrued as well. Bipedalism was also intimately connected to the evolution of the brain. The development of such an upright posture included a considerable degree of flexion at the base of the skull, lest the eyes face toward the sky.[5] Flexion at the base of the brain gave our earliest bipedal ancestors not only a forward-looking face, like that of the Taung child; it also resulted in a greater arching of what is now the top of the skull, consequently rounding and expanding the voluminous capacity of the braincase. Along the cranial base the brain bent over, twisting a bit more at the base, where the primitive neural functions lie, and stretching at the outer cerebral surface, where more advanced functions occur—where thinking takes place. A slightly expanded brain was thus an autocatalytic consequence of bipedalism.

This larger, modified brain, catalyzed by adaptations coincident with an upright posture, allowed even greater adaptability through behavioral plasticity—behavior guided somewhat more by thought and creation rather than by instinct. And so the autocatalysis of hominid brain evolution began in earnest. Those with larger brains were better able to procure food in a variety of environments as well as meet the challenges posed by predators. Indeed, the expanded part of the hominid brain allowed for learning, which in time could overshadow the diminished calling of instinctive behavior arising from deeper within the brain.

Thus one morphological step led intrinsically to another, and each was catalyzed by the advantages of niche expansion and population growth rather than forced by environmental change. It is true that if the environment had not changed and no savanna had been available, the process might have taken a different direction. Environmental change does shape evolution; it just does not catalyze it. But as it was, early hominids strode across the savanna just as pigeons now flock to cities; they were ready for it. Once the population found a way to grow, to expand its numbers into new environments, a positive feedback loop continued: more individuals meant more chances for novel genes to arise. Chance, the mother of invention, was allowed to procreate, and it just so happened that the variants it produced were launched in a savanna rather than a forest environment.

Populations of these creatures, later to be called australopithecines by their inventive descendants, spread across parts of the African landscape. Coincidentally, at this time the climate was getting cooler, the air was getting drier, and the forests were receding. It mattered little, for they were perfectly happy wherever they were. Well, maybe not *perfectly*—the occasional saber-tooth and the odd leopard dragged off a few of these comical creations into a cave, or dumped their stringy bodies next to a lake edge, hence our fossil record—but certainly they were adequately content, for they survived and gradually diverged from the cousins they left behind in the forest.

For some time, after their initial expansion onto the savanna, their populations were kept in check, and hominid evolution slowed—at least in the body parts seen in the fossil record. Somewhere, somehow, somebody's genes allowed teeth to grow a bit larger. The developmental time allotted for the incisor teeth in front became shorter, the growth of the grinding molar teeth quickened—not a huge genetic change, but one that toughened the molars for grinding savanna vegetation. It was a seemingly wondrous coincidence, and natural selection snatched the opportunity. When teeth grow, faces often do as well; they are intimately related in the developmental process.[6] Thus, in an instant of geological time, robust faces and large teeth spread throughout the population. The robust australopithecines were born.

There was just one problem. A large face gets in the way of an expanding brain. The simple fact that two objects cannot occupy the same space put a halt to increased cerebralization. One could not have a larger brain *and* a large robust face, at least not without a slew of other genetic and morphological changes, the likes of which nature has never seen. Nevertheless the robust australopithecines got what morphology they needed, and they managed to survive. For a while, anyway.

Somewhere else, somebody else, some other gene—and a different set of evolutionary exigencies found their way onto the African landscape. A chance mutation set the initial conditions for a different trajectory of evolution, leading to a different attractor of the natural selection process. We know not where, but a seemingly unfortunate group did not luck upon larger grinding teeth. Even a drying continent could not make them change. Perhaps while in search of food they longingly observed carnivores satiating their appetites on meat, or perhaps they just came upon a fresh steamy carcass. But for some small reason they tried a touch of meat and found it to their liking. In time a sharp stone flake, initially made by one of the clever new scavengers, helped to supplement their dental functions and cut meat from the carcasses that gave them sustenance. What's more, one did not need large grinding teeth to eat meat—the diverse set of teeth already in the mouths of these hominids did the job just fine. In fact a little reduction in molar size, and consequently in facial size, did not hurt their chances of survival one bit.

Quite the contrary. A bit of facial reduction allowed the frontal lobe of the brain to jut a little farther forward. Indeed the entire cranial contents had a bit more room to expand, but especially the neocortex—the thinking part. By little more than accident early *Homo* was born, and these augmented cerebral creatures began creating innumerable new ideas about how to survive. By happenstance they had overcome the morphological barriers encountered by their large-faced robust cousins. That set the autocatalytic wheels of evolution in motion: the brain expanded, their adaptability expanded, their population expanded, the genetic pool of variants increased, language began, and one of nature's most remarkable positive feedback loops took off. Those individuals who kept pace with the changes of their own species, and who fit into the new social regime, flourished and mated and passed along their genes, while others with lesser genetic endowments were less successful, and their genes got left behind.

The success of these early human animals, and the generalized adaptability of the species, allowed them to spread beyond their African home. They entered new lands, new environments. For some time their niche was limited to tropical and subtropical areas. But the variability of their large population eventually supplied the necessary biological means to cross the threshold into temperate regions. Their ability to push into new environments was augmented by the cultural adaptations afforded by the brain. Indeed their own inventions, their own culture, became their environment.

With its tremendous numbers spread across much of the Old World, this species was now poised for an influx of novel genes and evolutionary experiments. Many experiments no doubt failed. As it happened, apparently, a population in Africa first hit upon the combination of genes for the craniofacial traits we recognize today as being fully human. It was not the fragmentation of the species that led to the novelty but the continuous bombardment of new genes that comes with a large and widespread population. Those genes spread quickly (via gene flow) through a continuous population, aided by whatever selective value the new alleles conferred.

And the rest, as they say, is prehistory.

A Matter of Principles

Well, maybe it was not quite so simple. Innumerable events, big and small, certainly added moments of despair, and complication, to the picture I have painted. But the just-so story, albeit an indulgent fabrication, illustrates a number of principles that may well have shaped the evolution of our species from our earliest ancestors. I have no doubt that, as with computer models, the story contains at least a grain of truth, based on underlying processes, even if some details may later prove to be wrong as our explorations in the field begin to fill in the multitude of gaps in the fossil record.

At least four principles of autocatalytic evolution emerge from the above scenario. In brief, they are as follows:

- Each evolutionary change in an animal's size and shape may affect connected features of its morphology. Such morphological "conditions inherent in that which varies" may initiate further evolution in new directions. Our initial hominid origin, through bipedalism, was not so much a matter of locomotion as it was exploitation of a coincidental suite of advantageous features that tagged along.

- At the base of natural selection are members of one's own species. Irrespective of outside environmental changes, evolutionary novelties must first withstand the rigors of the developmental process, work in harmony with other adaptations to one's own body, and fit in with the social constraints of the group. Human cultural behavior entered the loop by lifting some selective restraints and imposing others.

- Large populations have more evolutionary opportunities for autocatalysis than small populations. Evolutionary success breeds more success and can quicken the pace of evolution.

- In biological evolution, invention is the mother of necessity, not vice versa. When a novelty born of chance coincides with opportunities in a neighboring environment, it can lead to the spread of the species.

None of these four points should surprise you, although you may join the many who find a degree of discomfort with one or another. So let's take them one at a time, dissect their component parts, and see if there is any evidence to support such a hypothesis of autocatalytic evolution.

Two Legs Good, Four Legs Bad

An animal's own parts must fit and function together. For example, the heart has to be big enough for the body size. One's legs, as Abe Lincoln was fond of noting, must be long enough to reach the ground. The heart, the legs, the body—it seems obvious that everything must hang together and function as a whole. But there is an interesting corollary of this principle. The need for coordination of body parts not only imposes limits that block evolution but also provides opportunities that help to launch new trajectories.

The importance of chance, coincidence, and chaos in evolution makes us *seem* very unlikely. But with the opportunities provided by autocatalytic evolution we are not as unlikely as it might appear—not in retrospect. Our very existence results from chance, but evoking "chance," you must recall, is just a fancy way of admitting we don't know the details of a chaotic system—we don't know all the initial conditions and all the deterministic equations. Sometimes those initial conditions, or at least elements of them, stare us right in the eyes and yet we miss them while looking too hard and not focusing properly. Sometimes, as Huxley put it, the variability on which natural selection acts is "determined in certain directions rather than others, by conditions inherent in that which varies."[7] The cards in the evolutionary deck were fortuitously rigged in our favor, making our peculiar evolutionary path somewhat more likely. The suite of features associated with bipedalism was a developmental and morphological attractor of states upon which natural selection could act.

The best example I can think of to illustrate how autocatalysis propels evolution in certain directions is the evolution of the human brain. One thinks that the uniqueness of humans lies in our brains, and to a large extent that is true. One then thinks that our remarkable brains need an equally remarkable explanation. To some extent they do. But for the most part the explanation of the evolutionary expansion of our brains is really rather simple.

Once an animal has a brain as sophisticated and evolved as that of *any* mammal, the human brain simply requires a few minor modifications. The basic capacities of thinking and doing, the very essence of cerebration, exist in all mammals. Our human brain is only marginally different from that of a chimp or even of a baboon. We just happen to think that our brains work better, because the thinking part—the neocortex—is expanded in humans.

The origin of the mammalian brain is undoubtedly a long, complex story. But the minor modifications needed to turn an apelike brain into a humanlike brain can be told in a very short story: it was a lucky coincidence that we became bipedal and later another stroke of luck that our faces reduced. And it was all autocatalyzed.

Our imaginary scenario of human evolution played upon the expansion of the brain for good reason. The evolutionary path was one of increased encephalization, a disproportionately greater development of the brain relative to the size of the body. Bizarre though it may seem, it may have been our upright posture that essentially caused our brains to expand. Our inherent upright condition, aside from whatever immediate selective benefits it gave us, set our longer evolutionary path on the trajectory of brain expansion.

The principle is simply a matter of how a brain shape emerges in development. As the human embryo grows, it folds. At the cranial end the neural system folds over on itself as well, forming a clump that becomes the brain. In a quadrupedal animal it folds a little bit. In a chimp it folds a bit more; otherwise the face would not face forward. And in a human the neural system folds a lot. This extreme folding scrunches the central, inner part of the bend into a tiny arch at the cranial base, giving it little space for future expansion—but that is where the primitive brain functions lie anyway. The same process of folding, however, expands the outer part. Coincidentally the neocortex of the cerebrum, that bit of gray matter with which we think, lies on the outer, expanding part.[8]

To illustrate the principle, just look at the book you are reading. As you open the book more and more, the spine becomes increasingly bent—it becomes more flexed. Simultaneously, the ends of the pages fan out into a greater and larger arch. Metaphorically speaking, the same physics of opening a book may have opened our minds.

This notion is based on the work of D'Arcy Thompson. Thompson was an independent thinker who almost single-handedly brought mathematics into the study of growth and morphology with his influential and poetic book, *On Growth and Form*, first published in 1917. As he wrote in his introduction, "Cell and tissue, shell and bone, leaf and flower, are so many portions of matter, and it is in obedience to the laws of physics that their particles have moved, molded and conformed."[9] Through the eloquent use of geometry, Thompson molded and conformed diagrams of human skulls into chimp skulls, and transformed chimp skulls into baboons. He did it, particle by particle, by reversing the key components of encephalization: he unflexed the cranial base and scrunched up the arching cranium on top (fig. 8.2).

Studies of comparative anatomy confirm a clear relationship between the volume of the braincase and the degree to which it is flexed at the base.[10] The more the brain folds at the base, the larger it tends to be. Quadrupedal animals tend to have little basicranial flexion, for the spinal cord enters from the back. Quadrupedal primates such as baboons have a bit more flexion and slightly larger brains, due in part to their upright feeding posture. A knuckle-walking chimp takes the trend further.

In *Australopithecus* the cranial base is folded more than that of a chimp and consequently can hold a bit more brain. It should come as no surprise; after all, the key components of the Taung child's skull that led Raymond Dart to place it in our ancestral lineage were the relatively forward-placed foramen magnum on a flexed cranial base and a slightly expanded cranium. Likewise Darwin noted that "a change in one part leads through the increased or decreased use of other parts, to other changes of a quite unexpected nature."[11]

Unexpected indeed. At least in part, our upright posture may have given us our bigger, better brains. It is impossible to say that bipedalism *caused* the development of a larger brain, but certainly the postural balancing of the head set some initial conditions for the arching of the cerebrum. One thing led to another, simply because of a basic developmental correlation. We owe our brains to geometric principles.

Other, physiological principles further augmented the connection between our upright posture and larger brain. Any brain is a sensitive organ that must be kept cool, and this generally requires preventing the body in which it resides from overheating. Whereas a giraffe's spots help to keep it cool, bipedalism alone helps to keep hominids cool, as noted by Peter Wheeler.[12] The basic principle is that upright beings out in the sun expose less of their body surface to direct sunlight than

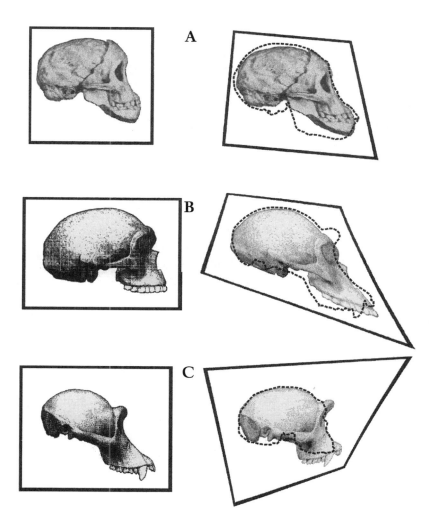

Figure 8.2 • Distortions of the skull, similar in style to those created by D'Arcy Thompson, show that much (but not all) of our cranial morphology can be attributed to simple shape transformations. **A.** A modest expansion of the cranial base transforms the Taung child toward the shape of a chimp (dashed line) of the same developmental age. **B.** An expansion and reorientation of the cranial base, along with facial expansion, transforms an Archaic *Homo sapiens* toward the shape of a modern chimpanzee. **C.** The opposite trends, basicranial and facial shortening, transform a chimpanzee into an Archaic *Homo sapiens*.

quadrupeds, which expose the entire length of the back from head to tail. The principle applies to plants as well as animals. If you look at trees that typically grow in the woods, you will see that their leaves spread out horizontally, gathering as much sunlight as possible. Trees growing in open areas, however, tend to have leaves that hang vertically so that the sun does not dry them out. Indeed this cooling principle is the evolutionary reason that blades of grass on the savanna are vertical and linear rather than horizontal and broad. This same coincidence of upright stance and physiological cooling aided the survival of hominids as they ventured out into the sun with their vulnerable brains atop their shoulders.

One cannot help wondering if there may be some counterintuitive implications of all this. Using hindsight we can trace marvelous feedback loops, including brain expansion, that began when a few coincidental changes tagged along with an upright stance. But natural selection could not have planned things that way; it has no foresight. Once a tree-dwelling primate gained access to resources on the ground, why should natural selection have upended the architecture any further? We became fully upright and bipedal for *some* reason—natural selection favored bipedalism over the exquisite gaits one gets on all fours. But bipedal walking is slow and clumsy. Other than freeing the hands to pick and gather fruit, which other primates can do anyway to a lesser extent, or to carry infants, who are too heavy to carry that way anyhow, there seems to be little selective value in our two-footed form of locomotion. There are many disadvantages, as we saw in the previous chapter. But perhaps, just perhaps now, the causes were reversed: natural selection favored our upright bipedal stance *because* of the coincidence of cerebral expansion, modest though it was at the beginning. Maybe bipedalism was just tagging along—not even as a benefit, perhaps, but a penalty worth paying. It's just an idea, but an intriguing one. It certainly should "get them talking"!

There is something unusual about modern humans, however. *Homo sapiens* do not fit the simple correlation between brain size and degree of flexion of the cranial base.[13] Our brains are much too big to fit the trend so well illustrated by other primates, including our australopithecine ancestors. Something else had to be involved in our latter-day brain enlargement. And that is where the trend of facial size reduction becomes so important. If the brain is to keep on growing, to keep on curling around, then the face has to get out of the way.

The face cannot change if natural selection favors those with a robust chewing apparatus, so tough luck for *Australopithecus robustus* populations—can't get there from here. After gaining some modest brain-power and becoming bipedal, they hit upon a novel adaptation that led them down the wrong garden path. But what a grand opportunity, what an autocatalytic chance, for early *Homo*, who hit upon *adaptability* rather than a dead-end adaptation.

With early *Homo* a new positive feedback loop was set in motion. A slightly reduced face allowed a slightly larger brain. (Coincidentally, the smaller face was a result of eating meat, for which large grinding teeth were not necessary; and meat provides rich resources for an active brain.) The larger brain allowed more adaptability, more awareness of how to find new sources of food. The larger brain, especially in newly expanded parts of the front and sides, also allowed more manipulative ability in the fingers and tongue. It just *happened* that the mammalian centers for those abilities were in the right place in the brain, making that trajectory of evolution more likely. The more agile hands and larger brains could now make tools and ultimately fire; tools and fire reduced the need for big strong teeth; smaller teeth allowed a smaller face; and the smaller face allowed a larger brain, completing the feedback loop.

Long after Huxley, once we knew about genes, it became clear that genes cannot mutate into something terribly convenient just because environmental circumstances dictate change. You can't always get what you want. We also knew that some gene mutations are more likely than others, because of their very nature. Genes, it cannot be denied, are "conditions inherent" among animals that vary. But because of developmental correlates of one body part with another, and developmental constraints of the same nature, evolution may take one course rather than another. Morphology is also a condition inherent in that which varies. I suspect that Thomas Huxley knew that, as did D'Arcy Thompson years later. And it is a coincidence of morphological correlation that set us feet first on the road of human encephalization.

All in the Family

My second point concerning the nature of autocatalytic evolution, or self-propelled evolutionary change, is that many evolved features have to do with the business of getting along with other members of one's own species. (The relative importance of this is represented in figure 5.2.) Again, this is not to say that adapting to one's habitat is a trivial concern. Of *course* any novel evolutionary product has to be consistent with survival in the outside world; for example, a polar bear would not stand a snowball's chance in hell of evolving in the tropics of Africa. But within the limits of features that are likely to evolve, given the exigencies of genes and the morphology that has come before, an individual with new variants must first play to the most important part of the environment: one's own species.

The evolutionary importance of one's own species is easy to see. At birth mammals must receive nurturing. This usually comes from the offspring's mother (to varying degrees) but often directly or indirectly from other species members as well. Being able to elicit this nurturing and benefit from it requires appropriate feeding behavior and morphology. One cannot survive without such infantile behavior and features, but some of those traits may be unrelated to adult adaptations. Sometimes, for instance, a newborn must compete with siblings for nurture, a rivalry illustrated in its most horrendous form by the black eagle chicks who fight to the death. Alternatively, occasional sibling *cooperation* might not be out of the question. Play among young siblings is often important for the learning process. And adult African bee-eaters, for instance, will sometimes help feed a sibling's or other close relative's chicks.

Among social mammals, which have evolved many competitive as well as cooperative social behaviors, a developing individual must somehow fit into the norms of behavior in order to benefit from existing mechanisms of group survival and also to find a mate. The relative importance of these social factors may vary, but they are inevitably important to some degree in determining Darwinian fitness (which in mammals depends primarily on mating success).

None of this is contrary to Darwinism. It was Darwin himself who came up with one of the very best examples of fitting in with one's own species: sexual selection. As the phrase implies, sexual selection is similar to (and in fact part of) natural selection. Males and females of a species must somehow choose one another for mating, and those who have traits making them more likely to be sexually chosen (attractive appearance, odor, behavior) are more likely to mate, produce offspring, and thus pass on their sexually attractive traits. Sexual selection is clearly an important aspect of natural selection. After all, Darwinian fitness has everything to do with how many offspring one leaves behind. However, survival in the wild is one thing; finding a mate and reproducing is another, at least among mammals such as humans and baboons. Traits useful for survival in the wild, such as blending in with the scenery and not making a sound that a predator might hear, may not be particularly helpful in attracting a mate. Undoubtedly the same is true for many animals; even cockroaches evolved special sounds and smells for mate attraction.

Sexual selection, in short, involves natural selection for features that attract mates. Peacock tail feathers and giant elk antlers are two classic examples. In both cases massive and in fact cumbersome male features evolved so that the respective creatures could display those features or wield them against rivals and thus attract mates. Most species have at least a few bizarre characteristics that are things of beauty or signs of good health to a potential mate, be they colorings, body sizes and shapes (males are often larger than females of their species), or other features. Humans are not excluded,[14] although our clothing and variety of cultures complicate the role of biological features in our sexual selection.

I believe that sexual selection is a good example, a fine example indeed, of autocatalysis—of how, among the many selective forces, the nature of one's own species determines and even propels evolution in one direction or another. But there are other, more subtle examples. And to illustrate the principles involved, as they were so poignantly illustrated to me, I have a special individual to help: my own son.

When my first son was born, he taught me more about biology than years of instruction and reading. Babies are as fragile as they are demanding, and thus are the first line of action for natural selection. For my boy's health and survival we owe an enormous debt of gratitude to our own species, who provided a most accommodating selective environment.

As it turned out, my son must have inherited some of my genes for facial shape, for he was a chinless wonder just like his dad. Maybe not chinless, for all *Homo sapiens* have chins (unlike their *Homo erectus* ancestors), but his "weak" chin is the product of a small, not very robust facial architecture. Whereas the little face was remarkably cute on my baby boy, it presented him with tremendous difficulties after birth: he could not latch onto his mother's nipple to feed from the breast. He tried and tried, as his remaining instincts told him to do, but then cried of hunger and frustration when he had no luck sucking out any milk. His little mouth and tiny chin just could not do it.

Much of natural selection, perhaps even most of natural selection, takes place long before a human or any other creature reaches maturity. Indeed selective pressure is quite heavy in the first few months of life. Just think of the process of facial reduction, a trend that characterized human evolution since the australopithecines and was critical for the continued increases in human brain size. In the wild, before culture gave us bottles with long, easily mouthed nipples, a child such as mine would not have survived. I am quite certain that many children even today do not survive for that very reason—their faces were reduced too far to allow them to suck on a normal human female nipple. The lesson is this: whatever features evolve for the sake of adult adaptations *must be consistent* with every other phase of development. That is not as simple as it seems, so a few more examples will help illustrate the difficulties involved.

First, one more example from my baby boys. Their instincts tell them not only to try to suckle but also to grab on for dear life. Just touch a baby's palm and it clenches with all its might. It is an evolutionary retention from our primate ancestors, much like the behavior of infant baboons I watch at Taung and Makapansgat, which grab the mother's underbelly hair so that she keeps all four limbs free for walking or running away. Most human infants can grasp your fingers with theirs tightly enough to be lifted right up, holding on with the same reliability as an infant baboon. My boys, however, would have made pathetic baboons. They held on but could not be lifted up as the parenting books say infants should. It does not matter now, not in the urbanized selective environment in which my boys are growing up. But that selective environment is a *human* one. The only relevant part of the environment that determined their survival was their own species. Morphologically we are equipped to carry them if they cannot hang on, albeit not without sore backs and tired arms. And our culture, in the form of baby carriages and bottles with artificial nipples and countless other items, provided the environment in which they could survive.

When evolutionary biologists, and particularly paleontologists, look at past trends we tend to look at adult adaptations. The important, definitive fossils have almost always been adults. One notable exception among hominid finds was the Taung child skull, but then again its status as part of our lineage was most severely questioned due to its premature morphology. Nevertheless, natural selection takes place most often among the young; infant mortality in Taung today is around 15 percent, and there is a heavy toll among juveniles as well. The Taung child, long set aside in evolutionary studies because of its youth, might in the long run tell us more than the adult fossils about the conditions of human evolution.

It is not just humans, with their vast cultural alterations of the environment, who largely determine the path of their own selection. Earlier we looked at the evolution of the giraffe's long neck and pondered the question about what good such a neck does for a baby giraffe. After all, young giraffes can get nowhere near the leaves at the tops of trees. But the fact of the matter is that adult giraffes have fairly long legs as well. A suckling baby giraffe must be tall enough to reach its mother's teats. An infant too short, or too short-necked, would be selected against. Again, it is the necessity of being adapted morphologically and behaviorally to one's own species that sets the selective parameters. This simple principle allows some evolutionary directions while eliminating the possibility of many other directions.

No species evolves totally free of the external environment, however, not even humans. All animals' bodies clearly reflect this fact. One cannot easily take a fish out of water, or expect a human to survive long in the water. So forces shaping our biological makeup, although increasingly environment-free, must at least render us environment-friendly. Still, the most important part of one's environment, at least for mammals, is one's own species. Having the right features for suckling determines success or failure. Having the right stuff to attract a mate determines whether or not one's genes get passed to the next generation, and all this catalyzes evolution in one direction or another.

On Populations and Punctuations

Although chaos is at the heart of autocatalytic evolution, its whimsical nature presents a problem. Evolution often hits an impasse. It needs to solve multiple problems at once, and this is difficult at both the genetic and the morphological levels. If lots of different features have to change at once to adapt to new environmental contingencies, a chaotic system may not be able to organize itself quickly enough. On the other hand what is most fascinating about chaotic systems of a certain type is that they *do* often hit upon novel organizational schemes. But their transition from one scheme to another is not necessarily quick and easy.

Chaotic evolutionary systems can take a long time to get themselves organized. Let's look at a simple example. For the human brain to expand, the female pelvis had to be widened in order to provide enough space for the birth of a large head. At least that was part of the solution that early *Homo* happened upon. But is that likely to happen quickly or to take a long time? Let's simplify matters by assuming that only two gene mutations have to occur: one for a larger brain, the other for a wider maternal pelvis. If the mutation rate for each of these outcomes is 1 in 80,000 or so, then the chance of two new mutations happening in proper succession, one in a mother and the other in her child, is 1 in 6.4 billion. Of course some of these mutations may already exist in the population, thus increasing the chances considerably. But a wide pelvis does little good for a female who is trying to walk bipedally, so natural selection would tend to eliminate such genes. Likewise a gene for a large-headed infant cannot get passed on if the infant cannot pass through the birth canal. One reaches an evolutionary impasse, like the robust australopithecines with their large faces limiting the size of their brains.

Let us suppose that the two evolutionary events required at the genetic level eventually coincide. But if the bright young child meets a premature demise, like the child from Taung, then the remarkable coincidence of genes gets lost as well. An unfortunate encounter with a leopard at the wrong evolutionary moment and, oops, the course of evolution is stalled if not changed entirely. Alternatively the child could be one of the 6 in 1,000 humans with a heart defect. Or it could succumb to a primitive form of malaria if bitten by the wrong mosquito. Despite the potential advantages of his or her large brain, and despite having chanced to pass through the adequate birth canal of the mother's pelvis, many other fates may conspire to prevent the genes from being passed on.

There are varied ways to overcome these specific limitations. One is time. With enough time, and a touch of luck, the simultaneous circumstance of the large brain and large birth canal may come together. The other is population size. The more the merrier—more people, more genetic opportunities. It's that simple: time and/or population.

A third possibility is that another course of evolution takes over. Perhaps a large-brained child is born by forcing it out earlier in the gestation period, before the brain's development reaches such tremendous proportions. But that still involves chance. The child's survival then depends on the lucky coincidence of a parental (or societal) animal that can manage altricial young. Or—as seemed to be the case with the robust australopithecines—natural selection latches onto another feature such as large tooth and face size, thus abandoning further cerebral expansion forever.

A fourth possibility is that evolution grinds to a halt. Such stasis, when it occurs, is not so much an equilibrium (though it can appear to be) as it is a paradoxical fixation at a certain stage of impasse. The evolutionary forces are still at work, but they are working against extremely poor odds. It is what Stuart Kauffman has called a complexity crisis.[15] On account of too many coincidental events (or genes or whatever) at once, this impasse allows neither progression nor regression. Even if novel features arise, they may not be promoted; it must be remembered that natural selection is still busy maintaining functional systems throughout the body.

What is surprising, however, is that occasionally, especially with time or with large populations, two or three lucky events do come together and set evolution on an entirely new course. It is clear that when events do come together—mutations for two complementary genes, say, coding for two complementary morphologies—they can break the deadlock, take off in the population, and spread like wildfire. Change is swift and sure. It is classic chaos. And it is classic punctuated equilibrium.

For years I doubted the theory of punctuated equilibrium because it was so commonly couched in terms of environmental upheaval. In such catastrophic situations a population would suffer, not evolve, for necessity has never been the mother of invention in evolution. My doubts were accentuated by Darwin's adherence to the dictum, "Natura non facit saltum" (nature does not make leaps).[16] But even Darwin could be wrong, and I should have paid attention to the way Huxley distanced himself from the dictum.[17] Given the surprising mechanisms of autocatalytic evolution, "punctuations" to an evolutionary "equilibrium" can occur from randomly timed coincidences of genetic and behavioral opportunities. Just as Darwin discovered with his barnacles that geological change was not necessary to induce variation, the scientific community must now come to realize that occasional fast change needs no outside catalyst. It can be autocatalyzed, and indeed *must* be autocatalyzed.

Bipedalism and encephalization, although they loom large, are only part of the story of human evolution. Our brains did not develop into such clever organs by size changes alone, although encephalization gave us a good start. Fine-tuning was also required, and it had to be coordinated with the brain's support systems. Such innovations, where genetically controlled, must arise by chance. How does one increase those chances? By increasing the population size or by waiting long enough.

Australopithecus, evolving slowly, actually changed little in over a million years. Dart postulated that australopithecine intelligence adapted the early hominids to a variety of environments. Their small degree of encephalization probably did allow that, but they did not thrive. Aside from their modest brainpower, the niche expansion afforded by bipedalism also helped keep their populations up. But further change in the brain was slow, despite the advantages such change could have rendered, for population numbers were low. Given all that time, it is not too surprising that other favorable traits might appear and be promoted instead, such as large teeth. But with time at least one population of *Australopithecus* (in which the large-teeth mutation never occurred or never caught on) finally accrued some key changes in the brain, and *Homo* was born.

At some point before 2.5 million years ago, someone picked up a rock and hit it against another to make a blade from the sharp fragments. That is all the inventiveness the *Homo* brain could muster at the time. But if it enhanced survival, it promoted population growth; and if the population grew, so did the opportunities for new variants to arise—new features for the work of natural selection. *Homo* shaped its future course of evolution in stone.

With early *Homo* we see the spread of our progenitors across Africa, and perhaps beyond, as clear evidence of an expanding population. At the same time we see increased complexity of the brain as well; the keen eyes of Phillip Tobias noted the small but significant bumps on the brain that function today in processing language and in the association of sights and sounds.[18] The neocortex also became more convoluted in humans, further increasing the area of gray matter.[19] The increased population allowed natural selection the luxury to experiment with a richer menu of choices, and upon receiving mutations to make the brain work better, the descendants enjoyed still greater survival odds and further population growth. The brain was a key component to the autocatalysis of modern *Homo*.

How did modern *Homo sapiens* arise? By the relentless genetic factory of a human population spread across Africa, Europe, and Asia. This is the true advantage of being a generalist species. Why did modern features first arise in Africa? It was a matter of chance; something similar could have occurred just as easily in Europe or Asia. Why did the features spread? Natural selection seized upon the opportunities born of chance.

The Mother of Necessity

For several chapters now, I have often slipped in a favorite turn of phrase of mine: invention is the mother of necessity. I should like to offer an analogy as further explanation of what I mean by that.

During our years of excavation at Taung, beer was the drink of choice at the end of long days as the winter evenings set in. A good South African beer went down well as the Southern Cross came up in the evening sky. Only occasionally, particularly on Sunday afternoons following a morning's peregrinations along the escarpment, a bottle of wine at a quiet spot on the edge of the Kalahari seemed to help our spirits.

The rarity of wine in our pattern of consumption was partly the product of an inhibitory factor: the lack of a corkscrew. My ever resourceful students would always find a way to dig the cork out of the bottle with a knife, or perhaps to force the cork into the bottle, but the procedure was always slightly cumbersome and most inelegant. We otherwise ate well, so the additional fiber added to our diets by floating fragments of cork was most unnecessary.

For Christmas one year my sister-in-law bought me a new Swiss Army knife. My old knife had seen better days, having been through the ravages of Boy Scouts during my youth and fossil excavation during my supposed adulthood. The new knife, moreover, came with a corkscrew attachment. That corkscrew was to change our lifestyle in the field forever.

With the new knife conveniently strapped to my belt, I faced our next field season at Taung with renewed initiative. And the opening of wine bottles became much easier than the opening of new fossil deposits. Suddenly we found ourselves sipping wine rather than beer at sunset on those beautiful evenings, still unmatched anywhere I have been. A tiny bit of new technology, in the form of a corkscrew barely noticeable among our equipment, had changed our behavior. An "invention," or at least a chance addition to my kit of survival tools, had become the mother of the necessity of wine.

Invention had met with opportunity, at least in the particular environment in which we lived. The nearby Afrikaans town of Hartswater, a mere twenty-minute drive from Taung, was a place where I did my banking and much of the shopping for our team. Afrikaners, however, are not great lovers of wine. Hartswater, also at the edge of the Kalahari, grows its own grapes and produces its own wine, laced with sugar, much to the contentment of the local populace. It is not a product aimed at the wine connoisseur. The local market also sold fine wines, but in a peculiar manner. Premium aged wine, say a nine-year-old cabernet sauvignon, sold for the same price as a young wine of only two years. If it was the same vineyard, the same wine, but a different vintage, it would all go for the same price. Sometimes the older ones were actually marked down in price so they would sell. It was a wine buyer's paradise—superbly aged wine for cheap.

Armed with a highly functional corkscrew, and living within a world of cheap but good wine, we had mutated and flourished. Invention had indeed met with opportunity. Wine was now an option for our periods of relaxation and saw many a sunset and moonrise.

The point of my little parable is this: adaptation is little more than coincidence. Our new habits of food and drink consumption seemed well adapted to the local environment, but it all started with a corkscrew, not the need for a corkscrew. The same is true in biological evolution. Adaptation gives the appearance of forethought, of planning, of being directed, but it is not. If I may quote Thomas Morgan, the geneticist, one more time with feeling, you should see what I mean: "a variation having appeared, chanced to find a suitable environment." The mother of invention was chance (subjected to morphological and genetic initial conditions), and either the chance paid off or it did not.

In evolution one thing leads to another. It's not much different in modern society. I cannot now live without fax machines, personal computers, E-mail, food processors, microwave ovens, and so on, none of which existed when I was born. I did not bring them into being; they just happened to become part of the tool kit with which I selected to live my life. One thing led to another. Our ancestors had only the tools that their genes and morphology allowed them, but once something came into being, they snatched up the opportunity.

This brings up an interesting point about how our ancestors spread to new lands and adapted to new environments. How, for example, could archaic *Homo sapiens* adapt to winters in the temperate regions? If necessity is not the mother of invention, it seems unlikely that such adaptations could take place. But the spread of humans across the globe, or the spread of any mammalian species, works differently in the world of autocatalysis.

A population can remain confined to tropical zones for long periods of time, as has happened to most primates, essentially knocking at the door of neighboring environments. Time and time again they knock away, with each new variant, until they stumble upon the key. Such variants might be genetically controlled innovations of morphology or physiology that would allow members of a population to endure the cold of temperate regions. In a large, widespread population the relentless mother of invention is more likely to offer such opportunities and also to push the survivors into new regions through continued growth. This process gives the appearance of a population moving in and adapting, but under autocatalysis the move works the other way around—the adaptation allows entry.

Of course with humans, innovations could also be cultural. The capture of fire, the creation of clothing, and eventually the construction of shelter greatly accelerated the human spread into new regions. Life in the temperate regions then became a necessity. It was a product of the adaptability afforded by our human brains.

The Law of Higgledy-Piggledy

Is autocatalytic evolution a defensible model? That is the question we must ask. It fits the human evolutionary data, but so could other models. Despite the inexorable logic of autocatalytic evolution (as I see it anyway), it is worth looking at some criticisms.

A lot of my scientific colleagues sort of like the idea of autocatalysis but want to think about it some more. That is good. Others say that the idea is nothing new. That is good too: the idea of autocatalytic evolution is very old, and each of its component parts has been argued before. But the new synthesis of old ideas with chance, coincidence, and chaos, as well as the application to human evolution, is still startling to some.

One critic said autocatalytic evolution implies only "that lineages show evolutionary change because they do and they are the animals they are." This attempt to belittle the idea actually hit the nail on the head. Spot on! Evolution occurs when, as one colleague put it, a species happens to be in the right place at the right time. But there is more to it than that, hence the importance of Huxley's conditions inherent in that which varies. The features a species evolves start the evolutionary wheels in motion in particular directions, for particular genetic and morphological reasons. Yes, humans are what they are because of themselves—but we can single out one or a few chance features and coincidences as key to this unique and remarkable autocatalysis. And in principle we could do the same for other species, explaining their different trajectories through evolution.

Another critic called autocatalytic evolution "a catch-all term that seems to explain everything and thus nothing." I half agree. Autocatalytic evolution could explain everything—the evolutionary patterns we see, the timing of evolutionary events, the direction of evolution within a lineage, and perhaps much more. Is that nothing?

Probably the greatest problem some scholars have with autocatalytic evolution is the model's reliance on large populations to increase the chances of evolutionary change. There is an entrenched notion that new species arise from small populations. Indeed there is no doubt that sometimes they do. A separation of populations into small isolates can lead to random changes through genetic drift and in turn result in speciation. But large populations can evolve into new species as well. Moreover, autocatalytic evolution is not necessarily meant to explain divergent speciation as much as it is the origin of novel features and adaptations.[20] Adaptation does not necessarily come with speciation; indeed speciation is often not accompanied by much morphological change at all.[21] As Julian Huxley wrote of speciation, "a large fraction of it is in a sense an accident, a biological luxury, without bearing upon the major and continuing trends of the evolutionary process."[22]

These criticisms of evolutionary autocatalysis are certainly worth considering. If there were no debate in science, our knowledge would not move forward. To paraphrase D'Arcy Thompson, disproof will not come to the scientist's confusion but will come as a scientist's reward.[23] Indeed, given what autocatalysis predicts to be the likely fates of small and large evolving populations—and thus the likely crash in biodiversity with a burgeoning human population—disproof may come as some solace to me.

Besides, much worse was said of the ideas of Darwin, Dart, and many others. At least no one has called autocatalytic evolution the "law of higgledy-piggledy"—well, not yet and not in so many words.

There are still many questions left to be asked and answered. Indeed, autocatalytic evolution must be tested, and it must be looked at in many different lights. More fossils of our hominid ancestors and the animals with which they lived will provide tests with each site we excavate. Living animals, computer models, and studies of our genes can be used to test the ideas further.

Alternative models can, should, and will be proposed. But any model must accommodate the three mischievous and ubiquitous forces. Chance, it cannot be denied, is at the heart of evolution, for new alleles of genes arise only by chance, and they must chance to find a suitable environment. Coincidences of many morphologies and environments leading to the survival and reproduction of an organism must be explained as well, without slipping into attempts at explaining mere coincidences. And finally all models must take note of genetic and morphological sensitivities to initial conditions, as well as the simultaneous nonlinear action of natural selection on a multitude of competing factors; they must accommodate chaos. Autocatalytic evolution succeeds at weaving the triumvirate of chance, coincidence, and chaos into the fabric of evolution through natural selection.

THE THEORY of autocatalytic evolution poses the other members of one's species as the most important part of the environment. It moves out from there to successively less important features of the external environment. Moreover, it argues that outside environmental change cannot cause appropriate new genes to appear. Only the genes can do that, and it is a mere coincidence if a new gene is suited for the environment. It is the adaptation that leads to the environment, not vice versa. Once some new features open the door, natural selection then takes its course. The chances of evolving are further enhanced by large populations; thousands of butterflies flapping their wings together are more likely to stir the air than just one.

We will continue to debate how evolution works, and we will continue to come up with new ideas and test them every way we can. But the stakes of our being right or wrong happen to be high for our species and many others right now. This is not just an arcane debate about an esoteric subject with no relevance. Although new ideas can be fun to play with, new insights about evolution may also have an opportunity to be very useful—an opportunity to address the serious troubles imposed upon the earth by complex biological systems.

9

The Beginning and the End of Evolution

THE STUDY of human evolution is among the most intriguing of all sciences. We have a natural curiosity about ourselves and our past, engendered by the probing abilities of our large brains. Sometimes, however, the scientific debates about evolutionary theory and fossil evidence seem far removed from anything practical or meaningful for our lives today. But the lessons of human evolution are important, have a broad value, and may even help us set the course for our future survival.

I often think back to a scientific conference I attended at a hotel on the shores of Lake Malawi, in the heart of Africa. Occasionally my concentration on the proceedings flagged. Staring out the conference room window at the lake, I found it easy to get lost in thought while watching the adaptable vervet monkeys scamper across the beach. Fish eagles soared above the waters, eyeing the same rich resources that provided sustenance to the local fishermen in their dugout canoes. It did not escape my attention that the three 3- or 4-million-year-old lake, the deepest and most stable in Africa, has the world's greatest biodiversity of cichlid fishes. The diversity is usually explained in the standard way, with small populations becoming isolated in the lake and, in a pulse of speciation, diverging into the hundreds of recognized fish species.[1] But it is not unlikely that the variants and species were auto-catalyzed, nearly every one of them, during the long prehistory of stable lake environments. Nature may abound with examples of autocatalysis.

The conference cruised along and recaptured my attention despite the natural wonders outside. We were struggling to decipher fossil clues about how evolution works, or at least how it used to work when our early ancestors set the stage for the eventual evolution of *Homo sapiens*. Sitting around a large table for five days, we discussed and argued and thought, and changed our minds a lot. This was real science, at its best. But our deliberations came to no general conclusion other than that we all had a lot more to think about. Our young science has much left to learn about nature.

As I sat fidgeting once again, pondering autocatalysis versus climatically induced pulses, I put a picture of my little baby boy in front of me for inspiration. He had already taught me a lot about the human biological predicament and was daily teaching me more: confirming nature's ludicrous tolerance for the crying of human infants, adding the ridiculousness of teething pains, and bringing to my attention the peculiar human risk of ear infection due in part to the strange shape and position of our skulls—why didn't evolution *do* something about all these infantile disasters? How did our transition from wild nature to the kinder environment of human culture allow the accumulation of so many features that clearly would have been maladaptive in nature's wilder realms?

But rather than cover old ground again, I started to think up new questions as I looked at the picture of my growing boy. What would the future hold for him and the hundreds of thousands of human babies born with him? And what of their descendants: would human evolution continue? It was hard to imagine, in the nearly pristine environment of Malawi, that our relatively comfortable niche would ever change. But thinking back to other places I had lived—cities like Johannesburg and St. Louis and even the small town in Ohio where I grew up—I realized that we are rapidly remaking our own environment. Things are changing, and changing fast. The changes we are making now may divert the course of our own evolution, and they have certainly displaced if not threatened the lives of plants and animals that used to inhabit the land now taken by cities and towns.

It soon became abundantly clear to me that our debates about evolution were not merely academic. There we were, discussing the possible impact of environmental change on the pace and course of early hominid evolution while all around us (though not in the immediate vicinity) environmental change was occurring at unprecedented rates. All because of the success—or failure to think ahead—of one particular species: *Homo sapiens.*

The questions I began asking myself at the conference still haunt me. Applying our competing theories to the present and future, I come up with two totally different scenarios. According to those who posit that environmental change spurs evolution, perhaps even causing pulses of evolutionary change, we should expect a massive amount of evolution among the poor and downtrodden species that humans have displaced. After all, the turnover-pulse model envisions small isolated populations as the source of evolutionary novelty, a fountain of new species. Species with large populations or even tremendously huge populations such as humans, pigeons, or cockroaches would grind to an evolutionary halt as novel variants got lost in the vast sea of genes. We would stay the same, biologically speaking, while the rest of the world adapted to the environment we so courageously but aimlessly created.

On the other hand—and I find this to be a rather disturbing hand— if I am right about the slowness of evolution through autocatalysis, the unlikelihood of rapid adaptive response to environmental change, and the vulnerability of small populations, then the evolutionary saga of the future will be just the opposite. When confronted with the ultimatum of adapt or die, most species will have no option but to die. Under an autocatalytic model, species usually evolve slowly and go extinct quickly; so a plant or animal caught in an increasingly difficult struggle for existence is likely to lose. Perhaps disproof of my notions *would* be my reward, for the scenario engendered by the theory of autocatalytic evolution does not bode well for the future our children will see.

Under autocatalytic evolution, in stark contrast to the turnover-pulse model, our species would be poised for remarkable evolutionary change. With our vast numbers, variants are already everywhere, sputtering along at low frequencies in our large gene pool but ready to be swept up into the human population by natural selection. Human evolution may have only just begun.

In the cold reality of scientific logic, both scenarios are a little overblown and fantastic. But that is what imaginations are for. Surely my colleagues are duly concerned about diminishing habitats and the potential extinction of many species that inhabit our changing planet; even the most ardent punctuationalist does not envision a pulse of species origins capable of replacing our current losses. And conversely, knowing some limits of natural selection even among large populations, I don't think that we are going to evolve overnight. But if we dissect these opposing views of evolution and single out the essential components, we can establish some guidelines for the future.

Many people, scientists and laypersons alike, ask what the future of evolution holds for humans. It would be ludicrous to try to predict specific trends. This entire book extols the importance of chance, coincidence, and chaos in determining the course and timing of evolution, so any precise prediction is destined to be wrong or at least highly improbable. But we do know that evolution, as well as extinction, works as a culmination of basic principles. So first let's try to apply the principles of autocatalytic evolution to our own evolutionary prospects, with a bit of conjecture for fun, and then we'll get down to the more important business of applying those same principles to the rest of life on earth.

What Will Happen to Us?

The first thing we know about future human evolution is that natural selection will have plenty of variability to work with. Much of the genetic material for our biological future already exists in the varied populations spread across the planet. If we are all dying at random with respect to the genes we carry, then nothing will change; natural selection and human evolution will have ground to a halt. But I suspect that it is not so. Natural selection is relentless, even in a population our size—*especially* in a population our size.

Let me give a sample of evolutionary possibility. Human life expectancy is getting longer, at least in developed countries and in many other parts of the world as well. As we get used to living to older ages, we tend to delay our reproduction; I was thirty-six when my first child was born. In the old days, before civilization extended our lives, most people had little chance of living to age thirty-six or beyond, and those who survived infancy and childhood took advantage of their biologically determined reproductive potential at a much earlier age. Thus genes that promote or allow bodily dysfunctions in older age could have accumulated at random. For example, if someone had a genetic predisposition for a form of adult cancer, it would not have mattered. Long ago, people would have left their offspring behind and died anyway before the cancer had a chance to spread throughout their bodies.

Today the situation is very different for many populations. A gene that makes one more susceptible to cancer has the possibility of leading someone to death during the reproductive years. Those who carry the gene will thus be less likely to leave as many offspring as those who do not. That is natural selection, pure and simple. We do not know for sure that such natural selection is occurring among humans today, but it seems likely that we are not dying at random with respect to some of our genes. Thus we are evolving. It should not escape your attention that weeding out a cancer-causing gene in this manner is also an example of autocatalysis: changes in our own species's behavior, extending both our life span and our likely reproductive years, lead to changes in the selective value of genes controlling an independent part of our physiology. We are unwittingly shaping our own evolution.

This scenario of an existing gene undergoing natural selection is sufficient to begin building a case for Darwinian evolution among modern humans. One could think of many other examples as well: those of us who are genetically predisposed to be more resistant to the noxious pollution in our cities, or to the low-level radiation silently searing through us from our television sets, may have a distinct selective advantage. But these hypothetical examples of evolution in action are not much different from the story of the cockroaches in my kitchen. In each case nothing new evolves, and natural selection simply promotes the frequency of one variant over another. And the subtle changes in our genes, coding for body chemistry rather than visible features, certainly would not show up in the fossil record for paleoanthropologists of the distant future.

Can anything *new* evolve in our species? Many scientists would say no.[2] They are the ones who hold that evolutionary advances take place in small isolated populations. Only in such situations, they claim, can a new allele or combination of alleles have a chance of sweeping through the population—perhaps by natural selection, perhaps by genetic drift. In large populations, by contrast, new mutant alleles have no chance and are condemned forever to get lost or stay at extremely low levels, even if they have a high selective value.

Balderdash. My computer simulations indicated just the opposite. But then again, my largest breeding population was 10,000 individuals (and a simplified version of reality) as opposed to the much larger and very real population of today. Can the same principles apply to a population of 6 billion people, a large chunk of whom are of breeding age? We don't need another simulation to test the hypothesis. Simple mathematical logic can win this argument.

Let's compare a hypothetical human population of 6 billion with one of only 10,000. We'll give each the same parameters of change. Which population is more likely to evolve with a brand new mutant allele?

Both populations, large and small, would have plenty of mutations to work with. In a small population a single mutant allele could catch on quite quickly—that much is true. And it could come to predominate in the population by either natural selection or genetic drift. But if the gene is lost for any reason—an automobile accident kills the gene's carrier, the individual contracts malaria or has a lethal allele on another gene, or the person just does not reproduce—then it may be a long time before the same mutant allele reappears.

How long it takes until the *same* mutant allele appears is an interesting and complicated question. Mutation rates refer to known changes per gene, but there are many places on a gene where a mistake could be made. The same gene might mutate fairly often, but hardly ever in the same way twice. If we allow some simplifying assumptions, just to make a point—if we say that a mutation with a specific effect can take place anywhere on a gene, and if we stick to simple coding errors—then we can calculate a rough estimate of how often the exact same mutation would occur. For a population of 10,000, once a *specific* mutant allele was lost, it could be another 38,000 years on average before that same mutant allele arose again.[3] So whereas mutant alleles may have a reasonable chance of being swept up by a small population, those chances do not come very often.

In dramatic contrast, a specific mutant allele could reappear about every 23 days in a population of 6 billion. If it is lost today in America, it may appear again next week in China or the following month in Sweden. Furthermore these recurrent mutations will continually appear in other parts of the world *independently* of the first success-ful origin. If the new allele indeed has a selective advantage, it can and will spread from wherever it starts. This, I suspect, is what happened about 100,000 years ago when our species became fully modern; by chance the new features first appeared in Africa. The same could happen today.

Our large human population is constantly bombarded with muta-tions; chance, the mother of invention, is exceptionally active when it has large numbers on its side. And natural selection can persist and drive evolution more consistently in a large interbreeding popu-lation, because fewer new alleles are lost due to isolation and genetic drift. The increased genetic variation then gives natural selection more opportunity to act.[4] This means that our species may be set to evolve at an unprecedented rate. Mutations occur by chance, but a large population increases our chances of finding and keeping the good ones.

But this initial example deals with just one mutant allele coding for just one beneficial trait. Often the important evolutionary changes in a species involve more than one gene. It is when one gets new gene *combinations* that one's morphology or physiology really changes. This principle was nicely illustrated by Darwin's finches on the Galá-pagos Islands. They could evolve quickly because occasionally, rarely, one "species" would hybridize with another "species." This would result in a whole new configuration of features, such as in beak shape or behavior patterns, that allowed them to survive.[5] But the Galápagos Islands support only relatively small populations of Darwin's finches, so it may be easy for one gene to chance upon another that results in an evolutionary advance from their tandem operation. Can it happen in a population of nearly 6 billion people?

Once again let's take the example of new mutant alleles that do not yet exist in the human population. The chances of both appearing at once in the same individual, or even the chances of their happening in a succession of parent and offspring, are vanishingly small. It is likely that they would appear in separate places, at separate times. In a small population the problem of getting them together is essentially a prob-lem of timing. One would need to have both new mutants appear and/or be sustained within a few generations of each other. If each one appears only every 38,000 years or so, then unless one of the alleles took hold in the population (e.g., through genetic drift) and was ready for the arrival of the complementary gene, it would be unlikely that the two specific mutants would ever meet.

On the other hand can two independent mutants meet in a large population? It may seem unlikely in a population of 6 billion. It indeed *is* unlikely in a population of 6 billion. But it is more likely than in a small population because we no longer have to worry about the time factor or losing either allele to genetic drift, since a replica is likely to pop up somewhere else in another 23 days or so. A separation of 6,000 miles makes the meeting of two new genes unlikely. But a separation of 38,000 years in the smaller population makes it virtually impossible. And our large human population is now more mobile, and more interbred than ever before—that increases the chances a bit.

One must also remember that we have been looking at a specific combination of mutations. But many other combinations may be possible that would be just as beneficial although they would set evolution on a different course. A different butterfly may flap its wings. Any single event is unlikely, but *some* event is *highly* likely in the world of evolutionary chaos.

The interbreeding of humans from across the globe brings up another possibility. As with Darwin's finches, new traits may arise from new combinations of *existing* gene alleles. Who knows what advantage may arise when, say, a young man from Asia mates with a female from Africa? Our human population is now full of hybridized populations, yielding unprecedented gene combinations and thus opportunities for natural selection to work. Our global mobility does not stall evolution, as the turnover-pulse model would have it, but accelerates the chances of new genetic potentials.

Whether it be by new mutant alleles or by new combinations of existing alleles, we *are* poised to evolve.

One More Lesson from Alice

Almost invariably, when I see science-fiction "preconstructions" of how humans may appear in the future, our descendants are depicted as having big brainy heads with small faces and bodies. I hasten to remind you that due to the very nature of evolution through chance, coincidence, and chaos, such prognostications are little more than fantasy. Just as our peculiar evolutionary path was not inevitable, our future course is not guaranteed. The exact course of evolution, or of any chaotic process, cannot be predicted. All we know is that evolution will continue. To paraphrase Alice in Wonderland, I don't care much where I go, so long as I get somewhere.[6]

On the other hand there are some evolutionary principles, and they can make the typical science-fiction scenarios a shade more believable (or not): our future appearance may be extremely murky, shall we say, but not unfathomable. It is at least worth considering the principles and some *possible* trends—just for fun, mind you.

It is true that our heads are not perfectly balanced upon our shoulders. The foramen magnum, the big hole at the base of the skull, is not perfectly centered. Thus there is potential for a greater degree of flexion at the base of the cranium, and given the way the brain grows, that may mean still more arching around the top (as illustrated by the pages of an opening book). So brain expansion is possible from that perspective, just as it was when our ancestors first became bipedal. Of course we would have to do something about the parturition problems that come with a larger cranium. We do not need to widen the pelvis, for children can be born earlier, before their brains get too large, even if we have to do it artificially. Birth by cesarean section is already remarkably common, allowing all sorts of evolutionary possibilities for brain size to increase. We have already artificially changed the selective criteria shaping our species, so why not change them some more? Viva autocatalysis!

Our faces may also shrink a bit, giving the brain even more room to grow. That would mean sacrificing teeth. Although I occasionally enjoy chewing on a tough bit of dried meat, a South African treat known as biltong, I really do not need any sophisticated chewing apparatus. I could probably live quite sufficiently without any teeth at all: I love chocolate milkshakes and can get quite adequate nutrition from soft foods such as scrambled eggs and bread, not to mention tofu (artificial selective criteria indeed). Even my weak-chinned baby boys were quite adept at gumming pieces of cooked chicken and fish into little bits for swallowing, only occasionally using their eight tiny incisors. So teeth and large faces we can do without.

Oddly enough, if you take a human head and flex the cranial base a bit more, increase the brain size, and reduce the relative size of the face, you get something that looks just like a child's head. So really all we need to do, in order to fulfill our evolutionary prognostications, is change our developmental pattern a bit so that we maintain childlike cranial proportions into adulthood. That evolutionary potential is already well known: it is called neoteny.

Neotenic evolution has already played a substantial role in our past. You may convince yourself by going to the nearest zoo and observing baby chimpanzees. Their facial proportions look remarkably human, or at least considerably more humanlike than the faces of adult chimpanzees. This is why so many scientists of the 1920s thought the Taung child was but a chimp. Of course we did not evolve directly from chimpanzees, but one can imagine that the common ancestor of chimps and humans had little offspring that looked somewhat like a baby chimp—and somewhat like an adult human. So the trend of neoteny may continue.

We know that in some respects the evolutionary scenario as depicted in science fiction has already begun. We see signs of continued facial reduction in modern populations. More and more people never develop their third molars—they never get their wisdom teeth. As evident from the fossil record of *Homo sapiens*, strong chins and third molars predominated more in earlier times than today.[7] The simian fold on my palm may be a bit primitive, but I like to think that my weak-chinned face is on the proper evolutionary course.

The reduction of the face has all sorts of implications. This is especially true for autocatalytic evolution, in which one trend leads to another, perhaps unrelated, course of evolution. We know that our facial reduction has continued, with only a few exceptions, since the origin of the genus *Homo*. As the face reduced, for whatever reason, so did the snout. A reduced snout left less area for the membranes and nerve endings responsible for olfaction—the sense of smell. With fewer nerves to accommodate, the olfactory lobes of the brain also reduced in relative size. Unlike other animals with a keen sense of smell and a full-size olfactory lobe, humans have tiny olfactory bulbs in their brains and a pathetic sense of smell. (This autocatalytic chain of consequences, incidentally, also leaves more space in the skull for the expansion of other parts of the brain.) It will be interesting for scientists in the future to find out if our olfactory apparatus diminishes even more or whether a minimal sense of smell is something that natural selection will maintain because of its survival value even in a modern world—warning us away from spoiled food, for example.

What else might we not use and lose? That notion still arouses skepticism, but I believe it is very important to insist that it takes the force of natural selection to maintain features. Without natural selection, features automatically (or autocatalytically) reduce. One possible mechanism is the probable mutation effect—the accumulation of detrimental mutant alleles once natural selection relaxes, simply because bad new alleles (reducing structure or function) arise far more frequently than good alleles. To be quite honest with you, the PME has never garnered much support among my colleagues. Indeed the theory was pronounced "dead" a few years ago. But rumors of its death are premature, for we see the PME in action all around us. Given our human control over our own destinies, with clothing and shelter and medicine allowing genes to survive that natural selection might have weeded out, *many* mutants can now accumulate. And they do.

A semiretired professor of anatomy back at the University of the Witwatersrand, a remarkable scholar and teacher named Jack Allan, was fond of unknowingly invoking the probable mutation effect to make fun of the people in my home country. "The next generation of Americans," he said, "are going to be born without legs!" Having aroused my interest and attention, he continued, doing his best to hide the smirk on his face: "It's because they drive everywhere in their cars. They don't need to walk anymore. They drive round and round a parking lot until they are right next to the door rather than parking further away and walking a few extra meters."

Jack Allan, of course, was quite correct about American parking-lot behavior. I countered his prognostication of limb reduction by pointing out that it takes a leg to reach the floor pedals of a car. "They'll redesign the cars," he said, breaking into a smile and a chuckle.

Quite clearly Professor Allan took the argument a bit far. But in many more subtle ways our culture and technology have indeed reshaped our selective environment. We may maintain our legs, but with eyeglasses our population's eyes get worse. With dentures and processed foods our teeth can disappear or grow in all askew, to an orthodontist's delight. And with medicine, perhaps, our natural immunity gets worse. This latter point intrigues me, and once again some subtleties have been illustrated to me by my favorite teachers of evolutionary principles, my own baby boys.

All young children love to put things into their mouths; it's part of the learning process. The array of dangerous and unhealthy things that could easily go into an infant human's mouth is horrendously frightening to every parent: small hard swallowable objects, sharp cutting fragments, infested human debris, infected filth, poisonous plants and liquids, and the list goes on. I wondered out loud if *Australopithecus* children did this, and one of my colleagues suggested that maybe a bit of relaxed selection was at play: modern parents worry and watch over their children more, so those whose excessive curiosity would have killed them ages ago tend to survive today. (He unknowingly acknowledged a touch of behavioral PME.)

On the other hand maybe *Australopithecus* children, including the Taung child, taste-tested every object within reach. In the short run, for any particular child, the consequences may have been devastating. It is not out of the question to propose that the Taung child could have been ill from eating something septic before the weakened boy fell into his watery grave; we will never know. In the long run, however, there may have been an advantage to the species. Such behavior automatically invokes natural selection for a very strong immune system. Only those with the best physiological responses to disease would have survived. As long as the reproductive rate was high enough, the population size would not have declined. Losing children with a healthier curiosity than immune system may have been affordable in the grand scheme of natural selection.

This bit of speculation, with little more scientific grounding in fact than Jack Allan's proposal about American legs, leads to a specific point of more immediate relevance. What is happening to our immune systems now? Given that we have benign inoculations, antibiotics, antiseptics, and all sorts of medicinal treatments to deal with infectious diseases, might our naturally invoked immune responses lose some of their efficiency? Might the reduced pressures of natural selection allow new, weaker immune system variants to accumulate through the PME? One wonders, but at this stage of our knowledge one can only speculate.

Such speculations require a look at the other side of the immunity story, and an intriguing side of evolution. How are diseases evolving?

Down with Disease

How does the human population eradicate a disease? The principle is quite simple: we eliminate its habitat. Smallpox has been virtually eliminated from the planet through the process of inoculation. If every inoculated human being has an immune system readily prepared to kill the smallpox virus, then the human-specific virus has fewer places to live. Enough humans were inoculated to wipe out most of the habitat the virus was capable of finding. It did not evolve quickly as its population dwindled; it went virtually extinct.

The opposite is true, to an extent, for diseases that go unchecked. Malaria, a parasite, comes to mind. Massive programs to eliminate mosquitoes, the carriers of malaria from human to human, did well to keep the disease in check in Africa. Those programs, however, were not as persistent as the ones that eliminated smallpox. Enough malaria parasites survived to reproduce and grow into a substantially large population among their human hosts, spread by their mosquito vectors. The increasing numbers of malarial parasites now have increasingly resilient variants evolving, new strains that are resistant to all known treatments and cures. Malaria is now, once again, a very frightening prospect.

These two examples bring up an interesting evolutionary problem as well as a striking paradox for epidemiologists. Exterminating a viral or bacterial population to the very last organism may be a sure way to eliminate a disease, but we are not likely to achieve this very often. Those darn little things can reproduce by the millions within a single human host in just hours. Some of the offspring, given the numbers, may differ from the parent microbes through minor mutations. Moreover these microorganisms do not play by the same genetic rules as mammals: they steal bits of DNA from one another to enhance their ability to evolve. That is why there are always new strains of flu—viruses evolve. Now, we have inoculations for many of the known viruses, and we have antibiotics that target many of the bacteria. Thus our immune systems are prepared to kill off particular viruses, and we constantly bombard bacteria with antibiotics. But the microscopic organisms don't go extinct.

Sheer numbers prevent most viruses and bacteria from going extinct. We managed to kill off the smallpox virus, or so it seems for now, but others are replicating by the billions in humans around the globe in the time it takes to read this paragraph. That allows for a lot of variability.

What are we doing about this? We are fighting off some bugs with immunity boosters and killing off the known bacteria that cannot live in the presence of antibiotics. That still, however, leaves our bodies wide open for all the variants against which we have no immunity and for which no antibiotics are effective. We are, according to many epidemiologists, helping to breed superbugs.[8] Our antibiotics, by failing to kill off every last microbe, have become part of the selective process that promotes resistant strains. We must just hope that the limits of natural selection apply to viruses and bacteria as well.

Our tremendous population size exacerbates the problem. Essentially, in terms of ecological theory, humans are an open niche, an abundant source of sustenance for another living organism. It seems unlikely that a supercarnivore is going to evolve that will consume our flesh from the outside; saber-toothed cats and the like long since gave way to smaller, sleeker carnivorous models. But other organisms may eat us from the inside. If a virus or a bacterium can find 6 billion warm homes in which to live and reproduce its own billions of offspring, with endless potential for new variants, then it is poised for rapid evolution. Having found the coincidence of a large, growing niche, like the hominids who successfully foraged and scavenged in the expanding African savanna, these tiny life forms were given a grand opportunity to evolve.

I find that prospect frightening, even with our antibiotics and functional immune systems. As an optimist, however, I must tell you that there is a positive side to bacterial evolution. To a large extent our bodies rely on bacteria, particularly in the gastrointestinal tract, where colonies of beneficial bacteria help digest our foods and even produce vitamins for us. If a new species of bacterium evolved that was more beneficial to the health and longevity of human beings, it would undergo positive selection, for its hosts would not die off so easily. We could develop a superflora within ourselves to help us live more easily.

But the bacteria in our guts are not the only species we need to thrive; we cannot survive without many life forms on this planet.

It's the End of the World as We Know It—and I Feel Fine?

What happens to our own species is one matter, and certainly cause for concern, but what becomes of other species is another matter entirely. Due to the massive publicity these days about threatened and endangered species, one tends to think of many plants and animals as being prone to extinction. Despite the hype, the perception is justified. It is true that many species, when reduced to small populations, go extinct, and that many are doing so now on a regular basis, due largely to the evolutionary successes of *Homo sapiens.*

On the other hand extinctions have always been part of evolution. As some species die off, are there not new opportunities for other species to evolve? Our human-induced environmental change may make small populations go extinct, but does it not provide opportunities for others that *prefer* the new habitats we create?

Pigeons certainly like our cities. They were indeed preadapted for cities, using buildings in the way they used rocks and cliffs. Cockroaches were able to expand their habitats as well, from the caves of Africa to the artificial caves we build in every environment. There are now over 3,500 species of cockroaches, due to their autocatalytic successes. Some were preadapted to our homes and have quite successful populations spreading through our kitchens. Cockroaches, like pigeons and humans, may be poised for further success. There is plenty of variability on which natural selection can work to make new pigeon and cockroach species.

Domesticated plants and animals provide another example of evolution in a world dominated by *Homo sapiens.* Cows and pigs, corn and wheat have essentially adapted to us and the environment we created as well. Domesticated life forms have arisen and covered the farmlands of the planet in a truly punctuated event. With the accelerated selection of the human hand, and now in some cases with artificial genes, there have been tremendous new varieties and perhaps species, hopeful monsters all of them. By virtue of being friendly or edible, all domesticated life forms evolved in a human-dominated environment.

But few of those new plants and creatures would be able to survive in the wild. Our domesticated plants need the artificial insecticides, herbicides, and fertilizers we devise to maximize crop growth. Without a technological helping hand they would be eaten alive, outcompeted by weeds, or simply wither in the field. They need the artificial environment we created for them; we bred them that way. They are highly specialized life forms.

The other problem of course is that we bred each domesticated crop from a small population very quickly, selecting *out* any variants that may have come along. We have developed what is called a monoculture of farm crops. They are well adapted to us and the environment we created for them, but they are not *adaptable.* Monocultures have the same problems that small populations do: there is little genetic variability for natural (or artificial) selection to propel the species any further. Meanwhile creatures other than humans that may like the plants are free to evolve resistance to the things we use to try to kill them. So between the monoculture of plants and the evolving life forms arrayed against them in growing numbers across the planet, our crops are in danger. The great Irish potato famine, during which an entire country faltered after its overreliance on a specialized South American plant (the potato), is just one notable example of what can and will happen.

Many plants are not even adaptable to increasing human demands, as they do not produce enough food for our growing population. Hence we look for genetic variability in wild strains of rice and in the grasses of the Americas, from which early agricultural peoples originally bred corn and other grain crops. Up to now we humans have selected for very specific features that suited our needs; we selected the plants' preadaptations, but we have not really created anything terribly new. Now their evolutionary potential is gradually reaching its limit. Huxley suggested that natural selection has its limits, and we have seen why, at both the genetic and morphological levels. Artificial selection is severely limited as well.

The limits of artificial selection clearly show up in domesticated animals. Pet dogs are highly bred for their coats or builds just as plants were bred for taste or productivity. But variability in other features was lost along the way. Dogs, particularly highly bred dogs, we must admit, are fairly stupid. (Perhaps we should have domesticated baboons.) Many dogs have lost the wiles and even the morphology it takes to survive in nature.

Cattle, sheep, and pigs are also quite peculiar beasts due to artificial selection. Only the latter can I picture surviving in the wild, and even so natural selection would be brutal. It is impossible for me even to try to envision herds of domesticated cattle wandering the savannas of Africa, for they have little or no defense and almost no adaptability. And, like monoculture plants, cattle highly bred for beef production have not necessarily maintained defenses from the predators that strike from within. Mad cow disease may be just the beginning of the end of these adapted but not adaptable beasts.

There is little doubt that *Homo sapiens* will continue to tinker with the genetics of plants and animals, and that the variants we select and promote will go through an evolutionary process of sorts. And it is wholly logical that some wild or semiwild species, adapted or adaptable to cities and other human environments, will evolve through a more natural form of selection. But there is also little doubt that many life forms will go extinct. And extinction will dominate creation if nothing else changes in the immediate future.

It's about Time, It's about Space

How does our human population eradicate a species? The principle is quite simple: we eliminate its habitat. Human encroachment on every part of the globe has squeezed out many a plant and innumerable animals. We tend to worry about the large ones like the rhino or the cuddly ones like the panda (well, despite their appearance, I would not risk cuddling a panda or any strong wild animal), but most of the extinctions we cause happen to living beings we never hear about and may never even have thought about. The large ones merely draw our attention to a much wider problem.

It is clear that humans have long been agents of extinction. In South Africa, for the past 3 million years or so, only about three or four species of large mammals went extinct every 100,000 years. Quite possibly the figure was greater, for there may have been more species arising and going extinct than we see in the fossil record. But it is clear from the fossil record that in just the past 10,000 years seven large mammal species have gone extinct. It is no mere coincidence that, with the recent burgeoning of human populations, large mammals went extinct much faster than before—about twenty times faster in South Africa (double the number of extinctions in a tenth of the time). South Africa got off easy; North America lost at least eight *genera* of large mammals, including as many as twenty-seven species, since *Homo sapiens* first entered the continent.[9]

Human hunting habits started the decline of these species' populations. Particularly in North America it appears that the hunting of mammoths and other large beasts drastically reduced their numbers and put them on the verge of extinction, even after they had survived numerous bouts of climatic and other environmental turmoil for hundreds of thousands of years. Whether human hands killed the last one or not does not matter—small populations are always in jeopardy of going extinct.

Yet only part of the problem has been hunting. We no longer need to hunt and kill animals directly; our effect can be just as devastating without traps and guns. All we have to do is take over the habitats of wild animals. Without resources to live upon, animals and plants as well die and eventually go extinct. The first to go are the ones with small populations and the specialists, of course. But even species that have thrived in large numbers, been through the ringer of variability selection, and not meandered off into specializations seem to be threatened. Why? Space to live is at a premium on this planet.

I often ask my students to guess how much space is available on this planet for each human being. We are not aquatic organisms for the most part, so I ask them to think only about the land surface area of the earth. If we were to spread humans evenly across the earth, take them out of the cities and give them their own little plot of land, be it forest, farm, city, desert, savanna, or tundra, how much space would each individual have?

Some students guess about one square mile each. No, it is less, I tell them. Then there is always the one student who guesses another extreme of one square meter. If that were true we would already be standing pretty much shoulder to shoulder, so that is not the answer either. My students and I gradually work our way down from the higher guesses to one square kilometer, to half a square kilometer, to even less. The truth is, using the round figure of 6 billion our population reached in 1999, that each individual would have an area roughly 160 meters by 160 meters. That is about the area of a small sports stadium.

Now, if we spread everybody out across the planet into their patches of land, imagine how you would survive. Even if you got a *good* patch of land, life would be a touch difficult. Within that limited space you would have to find your water, grow your food, gather your resources for housing and energy, dispose of your garbage, and park your car. You may even want to leave a little space for a road system so that others could drive their cars across these patches of land. I would certainly have to weigh the situation carefully before committing any land to growing tobacco for my pipe.

But now comes the rub. Each of us has to share the land with the other life forms on this planet.

Isn't it quite clear that as the human population grows—and it is still growing at a rapid enough rate to shoot us well beyond 6 billion people—space becomes more important? Other life forms will be squeezed out of existence. How many of us want to share our 160 by 160 meter space with Darwin's growing herd of elephants, or with a giraffe and a leopard? How many would take care of the diverse tropical flora at the expense of our own food crops? Thus animals and plants are forced into extinction as we take away their space for our own human purposes.

There is no doubt that plants and animals are going extinct every day as we encroach upon their habitats, but their rate of extinction is a bit difficult to discern. Part of the problem is that we do not know how many species exist. There are about 1.5 million *known* species of plants and animals. Although scientists have done very well at documenting species variability among certain kinds of critters—indeed I'd like to meet those who identified over 3,500 species of cockroaches—we have not yet documented the full range of life on earth. But we know that the true number of species is much greater than that of the known. For our purposes here it would not be unreasonable to use the common estimate of 12 million species of plants and animals. The rate at which these species go extinct also has to be estimated. Many attempts have been made to discern this rate, and it appears that a conservative estimate would put the extinction rate at about 5 percent of species per decade.[10]

When one plays with the numbers of 12 million species going extinct at 5 percent per decade, a fairly frightening picture emerges. That translates into 164 extinctions every day. If it were true, then a number greater than that of all *known* species (1.5 million) would be extinct in just over 25 years! *Homo sapiens*, incidentally, qualifies as a known species. If we kept up the rate of 164 losses per day, all 12 million species would be extinct in just over 200 years.

Let's dissect these figures a bit further. Many of the species going extinct are clearly just small populations of tropical plants or animals that may have gone extinct anyway. One must remember that there is nothing unnatural about extinction. Just as death must come with life, extinction must come with evolution. My heartfelt sentiments go out to the rhinoceros as it goes extinct, but it (or its closely related predecessors) has been around for a long time. We must accept *some* extinction.

When determining extinction rates, one must also consider taxonomic biases. Perhaps an observed extinction was not of a distinct species but just of a variant that an overzealous taxonomist classified incorrectly into species status. Whereas there may have been a regrettable loss of important genetic variability, we cannot record this as an extinction per se.

And new species may be evolving every day to replace some that have gone extinct. Evolution is a continuous process, so it would be naïve to think that it has stopped now. We have already looked at the evolutionary potentials of bird and insect species that rather like our urban environments. The same may apply to the rodents and other beasts that inhabit our cities. Certainly some plants have found a niche amid the concrete jungle, and plants and animals in more natural environments must be continuing the evolutionary process as well.

But one should still be worried. Despite any and all conceivable compensating factors, there is a net loss of species that has been accelerated by the encroachment of the large-brained biped into environments around the globe. That encroachment is getting greater on a daily basis. Even if the consequence were a net loss of only one species every day, the attrition would be far too high. With a steady rate of just one more extinction than creation per day, all life forms would be extinct in less than 33,000 years.

Now 33,000 years may seem like a long time to you. Perhaps you can sleep more easily tonight. But it is a mere pittance of evolutionary time. Using our mental clock of one second representing each year, 33,000 years is only a little over nine hours. That is a fraction of the time since *Australopithecus* roamed the African continent, and just about a third of the time that our own species has existed in its modern form.

I always encourage debate and inquiry among my students, and they are quick to latch onto this topic, for they do not want to believe what they are hearing. Often they say, "So what, we only need our own species and those that support us to survive. So as long as we have *Homo sapiens* and cattle and corn and a few other species, there is really no problem." I counter with my argument about monocultures and the need for variability. Besides, the air we breathe gets its oxygen from plants. "But most oxygen comes from life on the surface of the ocean, and the ocean itself is a source of food you have not accounted for." What they say is true about the oxygen, although I would hate to predict what kind of atmosphere we would have without the tropical rain forests of South America. But in terms of food production the ocean is essentially a desert. It is already overexploited and polluted, leading to the decline of many fish populations, the extinction of some, and severe problems for untold numbers of other organisms such as shellfish.

Besides, I say, the estimate of 33,000 years until every species on earth goes extinct is based on a steady rate of extinction. But it is known that ecosystems are more productive and more sustainable with greater biodiversity. The web of life is a complex and chaotic system that needs many ecological interactions. Once we lose a few component parts of a local ecosystem, it begins to falter. So with just a few extinctions, ecosystems begin to break down, become less sustainable, and the extinction rate accelerates. One extinction per day becomes two, and soon the earth reaches 164 *net* extinctions a day or more. And we don't even know which one might crash the whole system. Like the loss of an entire computerized population from the loss of a single female, the extinction of a single species could have a butterfly effect more devastating than the worst of hurricanes. We are the key parameter of the equations that may lead to a blue sky catastrophe. So any way you look at it, we are in trouble.

Students, and many people in general, still do not want to believe that we are destroying our life-support system. So I often use another analogy in which the tables are turned.

Imagine a highly successful species, a virus or bacteria perhaps, that is superbly adaptable. It could adapt to life in any mammal, be it human or cow, and eventually take over and kill it. The more successful it was, the more animals it would kill. Just think if HIV, the virus that causes AIDS, could continue to adapt to its human hosts and be spread by air or other means, like the flu, without the need for sexual transmission. It would eventually wipe out most or all of mammalian life, and upon killing all its hosts would eventually die itself.

Like an HIV virus gone wild, we humans are gradually but effectively destroying our host. Perhaps in evolution we got too much of what we wanted. With unprecedented success in reproduction and distribution, our human population is just getting too large. At our current rate of population growth how long would it be, theoretically speaking, before we were standing shoulder to shoulder with only one square meter each? Under 600 years. We would not need to wait 33,000 years for all other life forms to go extinct; the process would be accelerated by our own population growth.

To be sure, all life will not vanish from the planet. Some life forms will endure, and new types of ecosystems might evolve. This would, however, take a very long time, hundreds of thousands if not millions of years, before a diverse set of beings once again populated the planet. Evolution has to wait for chance, coincidence, and chaos to oblige. Whether or not humans would fit into a newly evolved regime is another question, and it may be unanswerable. Our survival would be a matter of chance.

Clearly our population must stop growing. It is already too large. "But surely our species is doing all right now, and there is more room to grow," say my students. Some of them, however, who are lucky enough to travel with me to the village of Buxton, get a painful lesson about how much room we humans have to grow. We go there for the fossil riches, but there is little else in the Taung region at the edge of the Kalahari desert. When the people of Buxton each got their 160 by 160 meter plot of land, there was little water, almost no arable land, few trees for housing and fuel, only hard earth in which to dispose of garbage, and no cars to park anyway. And the people of Buxton starve. The infant mortality rate is high, and disease takes its toll.

The Buxton community has often relied on me to lend digging equipment for funerals. Young and old die. My workers know I have a soft spot for children, and often I have given them loans or donations from my pocket when a new relative was born. It has saddened me too often when, weeks later, the funeral would be for the newborn.

Such is life in impoverished areas around the human-inhabited earth. There are people starving, people without jobs, people without homes, and people dying. Our planet cannot now support 6 billion people adequately, let alone the other life forms with whom we share our small globe. Ironically, the human population keeps growing, and the situation becomes even more tenuous. Is there a solution to this problem?

A Numbers Game

In preparing to write this book I read a number of other popular science books. I enjoyed every one, to greater or lesser degrees. Many of them, however, disturbed me in their final chapters. The authors, observing the human and planetary predicament as I have done in this final chapter, declare a solution and begin proselytizing to cajole the reader into accepting a particular plan of action. At the risk of making the same mistake, I should just like to point out two simple options for our population problem.

Populations grow when the birthrate exceeds the death rate. It is very simple mathematics, nothing more or less than a numbers game. Malthus and Darwin both knew the principles of population growth and built their theories upon those principles. Birthrates and death rates are the only parameters we need to deal with. So the solution to the human population predicament, and the consequent predicaments for other life forms, is one of the following: either we increase our death rate, or we decrease our birthrate.

My persistent students note that nature will probably take care of us. Through viruses that are rapidly evolving—an HIV virus takes to the air, an already airborne Ebola virus breaks out across Africa and beyond—the human population will be pruned. Alternatively, an aggressive and selfish *Homo sapiens* will continue to kill its own members through wars. And my students note further, unknowingly rebuilding a Malthusian theory, that many people will starve. Nature will take care of us with the Darwinian struggle for existence.

Perhaps they are right. As Malthus said so long ago, war, famine, and pestilence will keep our population in check. But it never ceases to amaze me how many students calmly accept the option of mass devastation, including war, to prune our population. Some cite Raymond Dart and claim that we are just following our natural human tendency to kill. But whether or not we are innately aggressive, as envisioned by Dart and others, is totally irrelevant. Our innate intelligence gives us a choice. We are, after all, a very *adaptable* animal.

But what can we choose? Do we want to increase the death rate? If war and famine is the way to go, I dread to think what kind of human we would be selecting for. It seems to me that there is a somewhat more pleasant solution, and it is achievable.

We could decrease the birthrate. We could have fewer, but happier and healthier, children. "But Dr. McKee" (I could see this question coming), "you have two children already. Aren't you contributing to the overpopulation problem?" Well, no, I'm quite happy with my two children for both personal reasons and grander ones. Two offspring is strictly the replacement rate: one to replace me, one to replace my wife Jean in the grand scheme of things. Some couples could have more, some could have fewer children, as long as it all worked out to replacement rate. The more who stick to a simple one or two, the better. Once we get even slightly beyond two children per couple, just as with Darwin's elephants or with the digital offspring in my computer, the population expands, and the Darwinian struggle for existence takes a turn for the worse.

How ironic that even with declining numbers of elephants in Africa we are willing to cull them in some areas to prevent the environmental degradation that results from their population growth. But so many among the most environmentally destructive species of all, *Homo sapiens*, are unwilling to curb our own population growth in a responsible manner.

It is just so simple: either the death rate goes up, or we take the responsibility to lower the birthrate. Surely the smartest, most adaptable mammal on earth can opt for the latter.

The Dawn of Humanity

Thoughtful men, once escaped from the blinding influences of traditional prejudice, will find in the lowly stock whence Man has sprung, the best evidence of the splendour of his capacities; and will discern in his long progress through the Past, a reasonable ground of faith in his attainment of a nobler Future.

—Thomas Henry Huxley,
Man's Place in Nature

On a weekend afternoon in the Makapansgat valley, I normally take the opportunity to walk quietly up the valley from our excavation site toward the rock pools hidden in the forest. Along the way it is hard to imagine the Darwinian struggle for existence. The same trees have been there for years, doing fine. Their roots are deeply sunk into the rich soil that long since eroded from the dolomite and quartzite of the valley walls. The younger trees and variety of smaller plants at their feet reach effortlessly toward the sky to share the generous supply of African sunlight. Day after day, the same types of birds and mammals feed from the forest's abundance of plant foods.

I often find a variety of particularly colorful spiders catching prey in their webs, always at the same convenient open spots along the path. Once I get to a particular bend in the stream, just down from the pools, there is always a small swarm of translucent, mothlike insects hovering above the reeds. It seems as if things never change, as if a perfect balance of nature has taken hold in the Makapansgat valley forever, with very little struggle at all.

No matter how the valley appears in a brief moment of time, evolution is continuing in its own subtle ways. A thousand years from now the valley will be changed, and even more so into the distant future. Everything I do in that valley has its own small effect on that future. Indeed, though we dream of winding back the clock and observing the evolution of life all over again, the chaotic evolutionary process begins again every day. We are part of the grand experiment.

As I stop along my walk to enjoy the scenery, I sit on a rock and look up at the valley wall. My eyes focus on the entrance to the Historic Cave of Mokopane, where thousands of his Ndebele people died in the horrible siege of 1854. The human struggle for existence, in its most contemptible form, once severely marred the peace here. Such travesties have probably happened more than once in the long history and prehistory of this magnificent valley, and sadly our *lack* of humanity toward our own species continues. Is this the way of a generalized, adaptable species? The dawn of true humanity, as we would like to think of it, has yet to come.

Wandering farther along the path through the heavily wooded area, I eventually make it to the pools. On more than one occasion I have just missed a troop of vervet monkeys, who scampered off into the trees with the sounds of my approach. Their fading chatter and a few bouncing branches from which they jumped are all that is left for me to observe. Nature's own remarkable genetic algorithm has made many adaptable species, monkeys among them.

As I sit by the water I try to imagine how *Australopithecus* would have lived in such an environment some 3 million years ago. Granted the valley was not as deeply etched by the incessant flow of water back then, but surely there was a valley with rock pools and teeming life. Perhaps, I muse, they occasionally reached into the water and grabbed one of the many crabs that slide down the slippery rocks underwater; certainly baboons do that today, so why not *Australopithecus*? And there is still a wide variety of native, edible plants growing all around the pools and up the hillside. Could it have been that much of a struggle to exist back then?

Sure, the occasional leopard or saber-toothed cat would also have wanted a sip of the cool water, and may have found a quick meal in the form of *Australopithecus*. And the early hominid did have to share the rich resources of the valley with baboons, and vervet monkeys, and many other beasts big and small that were after the berries and seeds and fruits. But the hominids survived and persisted as such for hundreds of thousands of years. What more could they have wanted? Why change if the struggle to exist was only a minor challenge?

It could only have been opportunity. The opportunity started in the *Australopithecus* genes and was selected by nature. Evolutionary change was not chosen. It just happened.

How different life was then from our lives today. Diving into the pools to cool myself, for a few moments I am nearly as vulnerable to nature as *Australopithecus*. But getting out, I have the advantages bestowed by 3 million years of evolution, biological and cultural, to take care of me. Our own human adaptability protects and ensures our continuance, at least for a while. Our adaptability and the way we use it, however, will profoundly shape the course of things to come.

Humans are just a transitional species poised to evolve, and with us other life forms will also evolve, and go extinct. But in sharp contrast to *Australopithecus* or any other species, we have opportunities we may *choose* to pursue. And we have liabilities that a sentient being must weigh, for the future of our support system hangs in the balance as well.

Evolution is riddled with chance, coincidence, and chaos, so there is no way to control our own evolution or that of other living beings. We cannot outsmart nature, even if we want to. We are merely the initial conditions of the future. But 6 billion butterflies can beat up quite a tempest. Somehow we must harness the profound and detrimental consequences our species has inflicted on the survival of others and allow nature to take its course.

Drying myself with my towel while the African sun does the rest of the job, I relax and ponder our future. Wouldn't it be nice if our genes allowed a better, even more adaptable brain, and then nature's selective criteria, unbeknownst to us, promoted such genes? Perhaps that brain would see our human predicament more clearly, with deference and respect to the many who have gone before. Our more intelligent descendants could aspire to correct the results of too much success without encouraging the worst, most violent elements of the Darwinian struggle. Every person could make a small contribution, a singular butterfly effect in accord with the whole of the species, any one of which might ensure that the future course of life on the planet operated with respect to the principles of nature as well as to the grander aspirations of a species once known as *Homo sapiens*.

Is it just an optimistic evolutionary fantasy? Stranger things have happened.

Notes

Chapter 1 • Chance, Coincidence, and Chaos

1. See Gould 1989.
2. Alternatively she could have been an anthropologist, or I could have become a lawyer.
3. T. H. Huxley (1854) cited in Desmond 1994, p. 191.
4. R. A. Fisher cited by J. Huxley 1958, p. 5.
5. Hesse [1943] 1987, p. 140.
6. Coincidentally, Lamarck published his *Zoological Philosophy* in 1809, in which he described a mode of evolution known as transformism.
7. Gregory 1949, p. 485.
8. From Lorenz 1979, cited in Gleick 1987, p. 322; Stewart 1989, p. 141; Gleick 1987, p. 8; Lewin 1992, p. 11; Kauffman 1993, p. 178; Adams 1989, p. 87.

Chapter 2 • Between a Rock and a Hard Place

1. The Rosetta stone is an ancient black stone with inscriptions in hieroglyphics, demotic characters, and Greek. It was the first key to deciphering Egyptian hieroglyphics.
2. Jordanova 1984.
3. T. H. Huxley [1863] 1901.
4. The fact of evolution was declared forcibly in 1942 by Julian Huxley, grandson of T. H. Huxley.
5. Kurt Gödel demonstrated the limits of mathematical proofs in 1931.
6. Conroy 1997, and various Glenn Conroy lectures.
7. Astronomer and mathematician John Herschel, cited in Mayr 1991, p. 49.
8. Cited in Desmond and Moore 1991, p. 492.

9. Darwin also thought that many had overstated the importance of natural selection and said as much in his final edition of *Origin of Species* in 1872.
10. T. H. Huxley [1878] 1888, pp. 306–307.
11. T. H. Huxley 1863, p. 148.
12. T. H. Huxley [1880] 1888, p. 312.

Chapter 3   •   *A Tale of Two Sites*

1. Thoreau [1854] 1986, p. 357.
2. Dart 1925a.
3. Keith 1925, p. 11.
4. Elliot Smith 1925, p. 235.
5. Darwin 1871, p. 199.
6. Jordanova 1984.
7. Kuhn 1970, p. 4.
8. According to the deductions of P. V. Tobias (1984).
9. Peabody's notes are held at University of California–Berkeley.
10. Dart 1926, p. 316.
11. Science does not always work by reason, and the methods can be irrational. See Kuhn 1970 and Feyerabend 1993.
12. See McKee 1993 for a full explanation of tufa formation.
13. Hrdlička 1925, p. 384.
14. More evidence is presented by McKee and Tobias 1994.
15. Dart 1926, p. 317.
16. Darwin 1871, p. 206.
17. Dart 1925b.
18. Dart 1953, p. 209.
19. T. H. Huxley [1870] 1894, p. 244.
20. Story as oft told by Bob Brain in his marvelous lectures.
21. Dart 1964, p. 52.

Chapter 4   •   *Speeding Up the Pace of Evolution*

1. Initially articulated by Eldredge and Gould 1972.
2. Just for the record, I prefer Beethoven.
3. Asfaw et al. 1999.
4. Sillen 1992; Lee-Thorpe et al. 1994.
5. Hughes and Tobias 1977.
6. Clarke and Howell 1972.
7. Thickening the soup, Wood and Collard (1999) suggest that most "early *Homo*" fossils should be placed in the genus *Australopithecus*. Along the same lines Asfaw et al. (1999) proposed the name *Australopithecus garhi* to represent 2.5-million-year-old fossils of a seemingly transitional species between *A. afarensis* and early *Homo*.
8. T. H. Huxley 1863, p. 119.
9. Darwin 1859, p. 48.
10. Vrba 1985, 1995.
11. Stanley 1998.
12. Boaz 1997.
13. Bishop 1994.
14. Behrensmeyer et al. 1997.

15. Just a subtle rejoinder to renowned evolutionary theorist John Maynard Smith, who in 1984 rather presumptuously welcomed paleontologists back to the "high table" of evolutionary studies.
16. "Species" are used as a measure of diversity for the simulations. Whether or not fossil "species" represent true biological species is another question.
17. The same was true of last appearances. McKee 1995.
18. I would like to note that Elisabeth Vrba has always been most gracious in this matter, as good scientists should be.
19. Darwin 1859, p. 463.
20. T. H. Huxley [1880] 1888, p. 322.
21. Browne 1995.
22. Darwin 1859, p. 82.
23. McCann and Yodzis 1994.
24. Walker 1984.
25. McKee 1991.
26. Brain and Sillen 1988.
27. Klein 1988.
28. Klein 1988.
29. Brain 1974.

Chapter 5 • *Rebels Without a Cause*

1. This tongue-in-cheek notion, originated by Sherwood L. Washburn, was cited by Wolpoff 1996, p. 109.
2. Hunt 1994.
3. Wheeler 1991, 1993.
4. Evidence from the genetic side is getting a bit more tenuous, e.g. Strauss 1999.
5. Van Valen 1973. Note that the Red Queen has since taken on a life of her own, and the hypothesis has since come to mean many different things to many different scientists. Here we will only deal with the relevant portions of the original hypothesis.
6. Lewis Carroll, incidentally, like Leigh Van Valen, was also very mathematically inclined. Carroll was the pen name for Charles L. Dodgson, an Oxford University mathematician.
7. Carroll [1897] 1992, p. 127.
8. Van Valen 1973, p. 21.
9. Carroll [1897] 1992, p. 13.
10. Vrba 1993.
11. One answer may be that there was no evolutionary stasis among the australopithecines. But whereas this may be true, whatever evolutionary change occurred prior to 2.5 million years ago was little in comparison to that which followed during the hominid divergence.
12. Potts 1996a, b.
13. Later apes, the chimps and gorillas, were considerably less adaptable and thus are currently threatened with extinction.
14. Indeed, landscape models have played a large role in evolutionary studies since Sewall Wright introduced the "adaptive topography" (see Wright 1931, 1977). But Wright dealt with peaks and valleys, not rivers.
15. Keep in mind that the analogy falls apart if taken further, for the river constantly pushes toward the sea. Evolution, on the other hand, has no destination.
16. Adapted from McKee 1999.
17. Carroll [1897] 1992, p. 28.

Chapter 6 • *The Mother of Invention*

1. Watson 1968.
2. Eyre-Walker and Keightley 1999.
3. Morgan 1910, p. 203.
4. Darwin [1858] 1993, p. 94.
5. MacFarland et al. 1974. Two have gone extinct in historic times due to the decimation of their populations, which began in the late 1600s when seafaring humans, naturalists and pirates alike, found them to be a good source of food. Goats and pigs and other introduced animals also play their part in keeping tortoise numbers low. The most vicious predatory creature alive—the tourist—currently threatens life on the island. Now eight of the remaining species or races (here we go with taxonomic problems again) are threatened with extinction. But that is another story.
6. Darwin 1839, p. 464.
7. Dawkins 1986.
8. See Kauffman 1993 for a technical example, or Kauffman 1995 for a more readable explanation.
9. Another class of genes of evolutionary interest are the "homeobox" genes. These are regulatory genes, in which mutations may have profound effects on morphology by drastically altering the developmental process. Such mutations would carry greater weight if included in a simulation such as the one described here. Indeed, they may eventually prove to be quite important in the origins of new species.
10. The rates used varied for different experiments. With too low a rate, very little change occurred within simulated populations. The simulations reported here would have yielded 2.5 mutations per individual across 100,000 genes. Eyre-Walker and Keightley (1999) estimate an average of 4.2 occurring since the hominid lineage separated from the chimp lineage.
11. For that matter, well over half of our genes have some equivalent in the earthworm.
12. Malthus [1798] 1826. Geometric increases are multiplicative, e.g. doubling from 2 to 4 to 8 to 16, etc., and thus can quickly rise to large numbers. This is as opposed to linear increases, which are additive and slow to accumulate (e.g. 1, 2, 3, 4, etc.).
13. Darwin 1859.
14. Darwin 1871, p. 213.
15. Lenski and Travisano 1994, Travisano et al. 1995.
16. Darwin [1858] 1993, p. 94.
17. See Felenstein 1971 and Lande 1994.
18. Kauffman 1993, 1995.
19. Browne 1995, p. 527.

Chapter 7 • *"You Can't Always Get What You Want . . ."*

1. Dobzhansky 1973, p. 125.
2. Krogman 1951, p. 54.
3. Darwin 1871.
4. Dawkins 1989.
5. Tobias 1987, 1994.
6. Laitman et al. 1992. It is as yet unclear exactly how much the cranial base must be flexed before the morphological correlate of laryngeal descent ensues; hence we know little about the ancestral position of the larynx.

7. There are many arguments about the anatomy of speech capabilities, particularly among Neandertals. Whereas there may be questions about when, and among whom, speech arose, the anatomical principles still apply.
8. McKee 1984.
9. This got through my "spell checker," or agent of selection. Does it get through yours?
10. Bromfield 1948, p. 79.
11. Darwin 1859, p. 137.
12. Jordanova 1984.
13. Weismann 1893.
14. Mitchell and Skinner 1993.
15. There is, however, a shorter cousin of the giraffe known as *Okapia johnstoni* in Congo.
16. Skinner and Smithers 1990.

Chapter 8 • Autocatalysis

1. Bielicki 1965.
2. Tobias 1981, 1994.
3. Kipling [1902] 1962.
4. The term *preadaptation* has been supplanted in some texts by the term *exaptation* on the basis of an argument by Gould and Vrba (1982). Whereas I opt for the former term, it does not imply any evolutionary forethought—just the chance nature of having features that are later of adaptive value in a different setting.
5. It is possible to adapt to bipedal posture differently, such as in birds, by bending the neck rather than the skull. Had this happened, our brain evolution may have taken a very different course. It would have been possible for brain expansion, like that of a dolphin, but the human brain was shaped by the initial condition of basicranial flexion and its anatomical correlates.
6. McCollum 1999. Note that the relationship between teeth and facial development means that the robust faces of East and South African australopithecines may have evolved more than once, both times as an autocatalytic response to excessive premolar and molar enlargement. See Wake 1991 for a discussion of the general principles involved.
7. T. H. Huxley [1878] 1888, p. 307.
8. The neocortex has been shown to be larger than expected in humans (for our brain size), as compared to other primates. Rilling and Insel 1999.
9. Thompson 1961, p. 7.
10. Ross and Ravosa 1993.
11. Darwin 1871, 151–152.
12. Wheeler 1991.
13. Ross and Henneberg 1991.
14. Lovejoy (1981) and Szalay and Costello (1991) made interesting attempts to tie sexual selection to the origin of bipedalism. The ideas are intriguing but difficult to test.
15. Kauffman 1993.
16. Darwin 1859, p. 194.
17. "Darwin's position might, we think, have been even stronger than it is if he had not embarrassed himself with the aphorism, '*Natura non facit saltum*,' which turns up so often in his pages. We believe that Nature does make jumps now and then, and a recognition of the fact is of no small importance." T. H. Huxley [1860] 1912, p. 77.

18. Tobias 1987, 1994.
19. Rilling and Insel 1999.
20. Darwin, in *On the Origin of Species*, barely dealt with the speciation process.
21. This point of view is taken by Ian Tattersall (1998), who then paradoxically goes back to speciation as the key element of evolution.
22. J. Huxley 1942, p. 389.
23. Thompson 1961, p. 10.

Chapter 9   •   *The Beginning and the End of Evolution*

1. Stiassny and Meyer 1999.
2. Tattersall 1998.
3. Based on nucleotide mutation rate data from Eyre-Walker and Keightley 1999 and assuming uniformity of the gene across the population. No account was taken of effective breeding size of the population, so a mutant could easily get lost if it arose in an individual who was past reproductive age. Changes in assumptions will change the actual figures used but will not alter the general point being made.
4. This is part of Fisher's fundamental theorem of natural selection. Fisher 1958.
5. Weiner 1994.
6. Carroll [1887] 1992, p. 51.
7. On the other hand agenesis (nondevelopment) of the third molar appears to have started with Asian *Homo erectus*.
8. E.g., Ewald 1994.
9. Klein 1992.
10. For a full discussion of varied estimates, see Stork 1997.

Literature Cited

Adams, D. 1989. *The Long Dark Tea-Time of the Soul.* London: Pan Books.

Asfaw, B., T. White, O. Lovejoy, B. Latimer, S. Simpson, and G. Suwa. 1999. *Australopithecus garhi:* A new species of early hominid from Ethiopia. *Science* 629–635.

Behrensmeyer, A. K., N. E. Todd, R. Potts, and G. E. McBrinn. 1997. Late Pliocene faunal turnover in the Turkana Basin, Kenya and Ethiopia. *Science* 278: 1589–1594.

Bielicki, T. 1965. The intensity of feedbacks between physical and cultural evolution. *International Social Science Journal* 17: 97–99.

Bishop, L. C. 1994. Pigs and the ancestors: Hominids, suids and environments during the Plio-Pleistocene of East Africa. Ph.D. dissertation, Yale University. Ann Arbor: UMI.

Boaz, N. T. 1997. *Eco Homo.* New York: Basic Books.

Brain, C. K. 1974. A hominid skull's revealing holes. *Natural History* 83: 44–45.

Brain, C. K., and A. Sillen. 1988. Evidence from the Swartkrans cave for the earliest use of fire. *Nature* 336: 464–466.

Bromfield, L. 1948. *Malabar Farm.* New York: Harper & Brothers.

Browne, J. 1995. *Charles Darwin: Voyaging.* London: Jonathon Cape.

Carroll, L. [1897] 1992. *Alice in Wonderland: Authoritative Texts of Alice's Adventures in Wonderland, Through the Looking Glass, The Hunting of the Snark,* ed. D. J. Gray. New York: W. W. Norton.

Clarke, R. J. 1988. A new *Australopithecus* cranium from Sterkfontein and its bearing on the ancestry of *Paranthropus.* In *Evolutionary History of the "Robust" Australopithecines,* ed. F. E. Grine, 285–292. New York: Aldine de Gruyter.

Clarke, R. J., and F. C. Howell. 1972. Affinities of the Swartkrans 847 hominid cranium. *American Journal of Physical Anthropology* 37: 319–335.

Conroy, G. C. 1997. *Reconstructing Human Origins—A Modern Synthesis.* New York: W. W. Norton.

Dart, R. A. 1925a. *Australopithecus africanus*: The man–ape of South Africa. *Nature* 2884 (115): 195–199.

Dart, R. A. 1925b. A note on Makapansgat: A site of early human occupation. *South African Journal of Science* 22: 454.

Dart, R. A. 1926. Taungs and its significance. *Natural History* 26: 315–327.

Dart, R. A. 1953. The predatory transition from ape to man. *International Anthropological Linguistic Review* I: 201–218.

Dart, R. A. 1964. The ecology of the South African man–apes. In *Ecological Studies in Southern Africa*, ed. D.H.S. Davis, 49–66. The Hague: Dr. W. Junk Publishers.

Darwin, C. 1839. *Journal of Researches into the Geology and Natural History of Various Countries Visited by the H.M.S. Beagle.* London: Henry Colburn.

Darwin, C. [1858] 1993. On the tendency of species to form varieties; and on the perpetuation of varieties and species by natural means of selection. In *The Portable Darwin*, ed. D. M. Porter and P. W. Graham, 89–104. New York: Penguin Books.

Darwin, C. 1859. *On the Origin of Species by Means of Natural Selection, or the Preservation of Favoured Races in the Struggle for Life.* London: John Murray.

Darwin, C. 1871. *The Descent of Man, and Selection in Relation to Sex.* London: John Murray.

Dawkins, R. 1986. *The Blind Watchmaker.* Harlow: Longman Scientific & Technical.

Dawkins, R. 1989. *The Selfish Gene.* 2nd ed. Oxford: Oxford University Press.

Desmond, A. 1994. *Huxley: The Devil's Disciple.* London: Michael Joseph.

Desmond, A., and J. Moore. 1991. *Darwin.* New York: Warner Books.

Dobzhansky, T. 1973. Nothing in biology makes sense except in the light of evolution. *American Biology Teacher* 35: 125–129.

Eldredge, N., and S. J. Gould. 1972. Punctuated equilibria: An alternative to phyletic gradualism. In *Models in Paleobiology*, ed. T.J.M. Schopf, 82–115. San Francisco: Freeman, Cooper, & Co.

Elliot Smith, G. 1925. The fossil anthropoid ape from Taungs. *Nature* 2885: 235.

Ewald, P. W. 1994. *Evolution of Infectious Disease.* Oxford: Oxford University Press.

Eyre-Walker, A., and P. D. Keightley. 1999. High genomic deleterious mutation rates in hominids. *Nature* 397: 344–347.

Felenstein, J. 1971. On the biological significance of the cost of gene substitution. *American Naturalist* 105: 1–11.

Feyerabend, P. 1993. *Against Method.* 3rd ed. London: Verso.

Fisher, R. A. 1958. *The Genetical Theory of Natural Selection.* 2nd ed. New York: Dover Publications.

Gleick, J. 1987. *Chaos: Making a New Science.* London: Cardinal.

Gödel, K. [1931] 1962. *On Formally Undecidable Propositions.* New York: Basic Books.

Gould, S. J. 1989. *Wonderful Life: The Burgess Shale and the Nature of History*. New York: W. W. Norton.

Gould, S. J., and E. S. Vrba. 1982. Exaptation—A missing term in the science of form. *Paleobiology* 8: 4–15.

Gregory, W. K. 1949. The bearing of the Australopithecinae upon the problem of man's place in nature. *American Journal of Physical Anthropology* 7: 485–512.

Hesse, H. [1943] 1987. *The Glass Bead Game*. London: Pan Books.

Hrdlička, A. 1925. The Taungs ape. *American Journal of Physical Anthropology* 8: 379–392.

Hughes, A. R., and P. V. Tobias. 1977. A fossil skull probably of the genus *Homo* from Sterkfontein, Transvaal. *Nature* 265: 310–312.

Hunt, K. 1994. The evolution of human bipedality: Ecology and functional anatomy. *Journal of Human Evolution* 26: 183–202.

Huxley, J. 1942. *Evolution: The Modern Synthesis*. London: George Allen & Unwin.

Huxley, J. 1958. The evolutionary process. In *Evolution as a Process*, ed. J. Huxley, 1–23. London: George Allen & Unwin.

Huxley, T. H. [1860] 1912. The origin of species. In *Darwiniana*. New York: D. Appleton & Co.

Huxley, T. H. 1863. *On Our Knowledge of the Causes of the Phenomena of Organic Nature*. London: Robert Hardwicke.

Huxley, T. H. [1863] 1901. *Man's Place in Nature and Other Anthropological Essays*. London: Macmillan and Co.

Huxley, T. H. [1870] 1894. Biogenesis and abiogenesis. In *Discourses Biological and Geological*. New York: D. Appleton & Co.

Huxley, T. H. [1878] 1888. Evolution in biology. In *Science and Culture and Other Essays*. London: Macmillan and Co.

Huxley, T. H. [1880] 1888. The coming of age of the "Origin of Species." In *Science and Culture and Other Essays*. London: Macmillan and Co.

Jordanova, L. J. 1984. *Lamarck*. Oxford: Oxford University Press.

Kauffman, S. A. 1993. *The Origins of Order: Self-Organization and Selection in Evolution*. New York: Oxford University Press.

Kauffman, S. A. 1995. *At Home in the Universe*. New York: Oxford University Press.

Keith, A. 1925. The Taungs skull. *Nature* 2905 (116): 11.

Kipling, R. [1902] 1962. *Just So Stories*. London: Macmillan and Co.

Klein, R. G. 1988. The causes of "robust" australopithecine extinction. In *Evolutionary History of the "Robust" Australopithecines*, ed. F. E. Grine, 499–505. New York: Aldine de Gruyter.

Klein, R. G. 1992. The impact of early people on the environment: the case of large mammal extinctions. In *Human Impact on the Environment: Ancient Roots, Current Challenges*, ed. J. E. Jacobsen and J. Firor, 13–34. Boulder: Westview Press.

Krogman, W. M. 1951. The scars of human evolution. *Scientific American* 185: 54–57.

Kuhn, T. S. 1970. *The Structure of Scientific Revolutions*. 2nd ed. Chicago: The University of Chicago Press.

Laitman, J. T., J. S. Reidenberg, and P. J. Gannon. 1992. Fossil skulls and hominid vocal tracts: New approaches to charting the evolution of human speech. In *Language Origin: A Multidisciplinary Approach*, ed. J. Wind, E. Chiarelli, B. Bichakjian, and A. Nocentini, 385–397. Dordrecht: Kluwer Academic Publishers.

Lande, R. 1994. Risk of population extinction from fixation of new deleterious mutations. *Evolution* 48: 1460–1469.

Lee-Thorpe, J. A., N. J. Van der Merwe, and C. K. Brain. 1994. Diet of *Australopithecus robustus* at Swartkrans from stable carbon isotope analysis. *Journal of Human Evolution* 27: 361–372.

Lenski, R. E., and M. Travisano. 1994. Dynamics of adaptation and diversification: A 10,000-generation experiment with bacterial populations. *Proceedings of the National Academy of Science* 91: 6808–6814.

Lewin, R. 1992. *Complexity: Life at the Edge of Chaos*. New York: Macmillan.

Lorenz, E. 1979. Predictability: Does the flap of a butterfly's wings in Brazil set off a tornado in Texas? Address to the American Association for the Advancement of Science, Washington, D.C., December 29, 1979.

Lovejoy, C. O. 1981. The origin of man. *Science* 211: 341–350.

MacFarland, C. G., J. Villa, and B. Toro. 1974. The Galápagos giant tortoises (*Geochelone elephantopus*)—Part I: Status of the surviving populations. *Biological Conservation* 6: 118–133.

McCann, K., and P. Yodzis. 1994. Nonlinear dynamics and population disappearances. *The American Naturalist* 144: 873–879.

McCollum, M. A. 1999. The robust australopithecine face: A morphogenetic perspective. *Science* 284: 301–305.

McKee, J. K. 1984. A genetic model of dental reduction through the probable mutation effect. *American Journal of Physical Anthropology* 65: 231–241.

McKee, J. K. 1991. Palaeo–ecology of the Sterkfontein hominids: A review and synthesis. *Palaeontologia Africana* 28: 41–51.

McKee, J. K. 1993. Formation and geomorphology of caves in calcareous tufas and implications for the study of the Taung fossil deposits. *Transactions of the Royal Society of South Africa* 48: 307–322.

McKee, J. K. 1995. Turnover patterns and species longevity of large mammals from the Late Pliocene and Pleistocene of southern Africa: A comparison of simulated and empirical data. *Journal of Theoretical Biology* 172: 141–147.

McKee, J. K. 1999. The autocatalytic nature of hominid evolution in African Plio–Pleistocene Environments. In *African Biogeography, Climate Change, and Early Human Evolution*, ed. T. G. Bromage and F. Schrenk, 57–67. New York: Oxford University Press.

McKee, J. K., and P. V. Tobias. 1994. Taung stratigraphy and taphonomy: Preliminary results based on the 1988–93 excavations. *South African Journal of Science* 90: 233–235.

Malthus, T. R. [1798] 1826. *An Essay on the Principle of Population*. 6th ed. London: Ward, Lock & Co.

Mayr, E. 1991. *One Long Argument: Charles Darwin and the Genesis of Modern Evolutionary Thought*. London: Penguin Books.

Mitchell, G., and J. D. Skinner. 1993. How giraffe adapt to their extraordinary shape. *Transactions of the Royal Society of South Africa* 48: 207–218.

Morgan, T. H. 1910. Chance or purpose in the origin and evolution of adaptation. *Science* 31: 201–210.

Poirier, F. E, and J. K. McKee. 1999. *Understanding Human Evolution*. 4th ed. Upper Saddle River: Prentice Hall.

Potts, R. 1996a. Evolution and climate variability. *Science* 273: 922–923.

Potts, R. 1996b. *Humanity's Descent—The Consequences of Ecological Instability*. New York: William Morrow & Co.

Rilling, J. K., and T. R. Insel. 1999. The primate neocortex in comparative perspective using magnetic resonance imaging. *Journal of Human Evolution*, in press.

Ross, C. F., and M. Henneberg. 1995. Basicranial flexion, relative brain size, and facial kyphosis in *Homo sapiens* and some fossil hominids. *American Journal of Physical Anthropology* 98: 575–594.

Ross, C. F., and M. J. Ravosa. 1993. Basicranial flexion, relative brain size, and facial kyphosis in nonhuman primates. *American Journal of Physical Anthropology* 91: 305–324.

Sillen, A. 1992. Strontium–calcium ratios (Sr/Ca) of *Australopithecus robustus* and associated fauna from Swartkrans. *Journal of Human Evolution* 23: 495–516.

Skinner, J. D., and R.H.N. Smithers. 1990. *The Mammals of the Southern African Subregion*. 2nd ed. Pretoria: University of Pretoria.

Stanley, S. M. 1998. *Children of the Ice Age: How a Global Catastrophe Allowed Humans to Evolve*. New York: W. H. Freeman.

Stewart, I. 1989. *Does God Play Dice? The New Mathematics of Chaos*. London: Penguin Books.

Stiassny, M.L.J., and A. Meyer. 1999. Cichlids of the Rift Lakes. *Scientific American* 280: 64–69.

Stork, N. E. 1997. Measuring global biodiversity and its decline. In *Biodiversity II*, ed. M. L. Reaka-Kudla, D. E. Wilson, and E. O. Wilson, 41–68. Washington: Joseph Henry Press.

Strauss, E. 1999. Can mitochondrial clocks keep time? *Science* 283: 1435–1438.

Szalay, F. S., and R. K. Costello. 1991. Evolution of permanent estrus displays in hominids. *Journal of Human Evolution* 20: 439–464.

Tattersall, I. 1998. *Becoming Human: Evolution and Human Uniqueness*. New York: Harcourt Brace.

Thompson, D. W. 1961. *On Growth and Form*. Cambridge: Cambridge University Press.

Thoreau, H. D. [1854] 1986. *Walden and Civil Disobedience*. New York: Penguin Classics.

Tobias, P. V. 1981. *The Evolution of the Human Brain, Intellect, and Spirit*. Adelaide: University of Adelaide.

Tobias, P. V. 1984. *Dart, Taung and the "Missing Link."* Johannesburg: Witwatersrand University Press.

Tobias, P. V. 1987. The brain of *Homo habilis*: A new level of organization in cerebral evolution. *Journal of Human Evolution* 16: 741–761.

Tobias, P. V. 1994. The craniocerebral interface in early hominids: Cerebral impressions, cranial thickening, paleoneurobiology, and a new hypothesis on encephalization. In *Integrative Paths to the Past: Paleoanthropological Advances in Honor of F. Clark Howell*, ed. R. S. Corruccini and R. L. Ciochon, 185–203. Englewood Cliffs: Prentice Hall.

Travisano, M., J. A. Mongold, A. F. Bennett, and R. E. Lenski. 1995. Experimental tests of the roles of adaptation, chance, and history in evolution. *Science* 267: 87–90.

Van Valen, L. 1973. A new evolutionary law. *Evolutionary Theory* 1: 1–30.

Vrba, E. S. 1985. Ecological and adaptive changes associated with early hominid evolution. In *Ancestors: The Hard Evidence*, ed. E. Delson, 63–71. New York: Alan R. Liss.

Vrba, E. S. 1993. Turnover-pulses, the Red Queen, and related topics. *American Journal of Science* 293-a: 418–452.

Vrba, E. S. 1995. The fossil record of African antelopes (Mammalia, Bovidae) in relation to human evolution and paleoclimate. In *Paleoclimate and Evolution*, ed. E. S. Vrba, G. H. Denton, T. C. Partridge, and L. H. Burkle, 385–424. New Haven: Yale University Press.

Wake, D. B. 1991. Homoplasy: The result of natural selection or evidence of design limitations? *American Naturalist* 138: 543–567.

Walker, A. 1984. Extinction in hominid evolution. In *Extinction*, ed. M. H. Nitecki, 119–152. Chicago: The University of Chicago Press.

Watson, J. D. 1968. *The Double Helix: A Personal Account of the Discovery of the Structure of DNA*. New York: Atheneum.

Weiner, J. 1994. *The Beak of the Finch*. London: Vintage.

Weismann, A. 1893. *The Germ-Plasm: A Theory of Heredity*. London: Walter Scott.

Wheeler, P. E. 1991. The influence of bipedalism on the energy and water budgets of early hominids. *Journal of Human Evolution* 20: 117–136.

Wheeler, P. E. 1993. The influence of stature and body form on hominid energy and water budgets; a comparison of *Australopithecus* and early *Homo* physiques. *Journal of Human Evolution* 24: 13–28.

Wolpoff, M. H. 1996. *Human Evolution, 1996–1997 Edition*. New York: McGraw Hill.

Wood, B., and M. Collard. 1999. The human genus. *Science* 284: 65–71.

Wright, S. 1931. Evolution in Mendelian populations. *Genetics* 16: 97–159.

Wright, S. 1977. *Evolution and the Genetics of Populations*. Vol. 3, *Experimental Results and Evolutionary Deductions*. Chicago: The University of Chicago Press.

Index

About the Author

JEFFREY K. MCKEE earned his Ph.D. at Washington University, St. Louis, in 1985. The following year he joined the academic staff of The University of the Witwatersrand in Johannesburg, South Africa. Under the direction of Phillip Tobias, he led excavations at the Taung fossil site for seven years. Later, Dr. McKee began further excavations at Makapansgat, the oldest fossil hominid site in South Africa. In 1996 he returned to his home state of Ohio, where he holds appointments in the Department of Anthropology and the Department of Evolution, Ecology, and Organismal Biology at The Ohio State University. He has published numerous academic papers and coauthored (with Frank E. Poirier) *Understanding Human Evolution* (4th ed.).